Amusement Park
of the Future

STEM Road Map
for Middle School

Grade
6

Amusement Park
of the Future

Grade
6

STEM Road Map
for Middle School

Edited by Carla C. Johnson, Janet B. Walton, and
Erin Peters-Burton

NSTApress

National Science Teachers Association

Arlington, Virginia

National Science Teachers Association

Claire Reinburg, Director
Rachel Ledbetter, Managing Editor
Deborah Siegel, Associate Editor
Donna Yudkin, Book Acquisitions Manager

ART AND DESIGN
Will Thomas Jr., Director, cover and
 interior design
Himabindu Bichali, Graphic Designer, interior
 design

PRINTING AND PRODUCTION
Catherine Lorrain, Director

NATIONAL SCIENCE TEACHERS ASSOCIATION
David L. Evans, Executive Director
David Beacom, Publisher

1840 Wilson Blvd., Arlington, VA 22201
www.nsta.org/store
For customer service inquiries, please call 800-277-5300.

NSTA is committed to publishing material that promotes the best in inquiry-based science education. However, conditions of actual use may vary, and the safety procedures and practices described in this book are intended to serve only as a guide. Additional precautionary measures may be required. NSTA and the authors do not warrant or represent that the procedures and practices in this book meet any safety code or standard of federal, state, or local regulations. NSTA and the authors disclaim any liability for personal injury or damage to property arising out of or relating to the use of this book, including any of the recommendations, instructions, or materials contained therein.

Library of Congress Cataloging-in-Publication Data
Names: Johnson, Carla C., 1969- editor. | Walton, Janet B., 1968- editor. | Peters-Burton, Erin E., editor. | National Science Teachers Association.
Title: Amusement park of the future, grade 6 : STEM road map for middle school / edited by Carla C. Johnson, Janet B. Walton, and Erin Peters-Burton.
Description: Arlington, VA : National Science Teachers Association, [2017] | Includes bibliographical references and index.
Identifiers: LCCN 2017022187 (print) | LCCN 2017023828 (ebook) | ISBN 9781681404844 (e-book) | ISBN 9781681404837 (print)
Subjects: LCSH: Tall buildings--Study and teaching (Middle school)--United States. | Skyscrapers--Study and teaching (Middle school)--United States.
Classification: LCC NA6230 (ebook) | LCC NA6230 .A48 2017 (print) | DDC 720/.483--dc23
LC record available at https://lccn.loc.gov/2017022187

The *Next Generation Science Standards* ("*NGSS*") were developed by twenty-six states, in collaboration with the National Research Council, the National Science Teachers Association and the American Association for the Advancement of Science in a process managed by Achieve, Inc. For more information go to www.nextgenscience.org.

CONTENTS

Part 1: The STEM Road Map: Background, Theory, and Practice

Part 2: Amusement Park of the Future: STEM Road Map Module

CONTENTS

ABOUT THE EDITORS AND AUTHORS

Dr. Carla C. Johnson is the associate dean for research, engagement, and global partnerships and a professor of science education at Purdue University's College of Education in West Lafayette, Indiana. Dr. Johnson serves as the director of research and evaluation for the Department of Defense–funded Army Educational Outreach Program (AEOP), a global portfolio of STEM education programs, competitions, and apprenticeships. She has been a leader in STEM education for the past decade, serving as the director of STEM Centers, editor of the *School Science and Mathematics* journal, and lead researcher for the evaluation of Tennessee's Race to the Top–funded STEM portfolio. Dr. Johnson has published over 100 articles, books, book chapters, and curriculum books focused on STEM education. She is a former science and social studies teacher and was the recipient of the 2013 Outstanding Science Teacher Educator of the Year award from the Association for Science Teacher Education (ASTE), the 2012 Award for Excellence in Integrating Science and Mathematics from the School Science and Mathematics Association (SSMA), the 2014 award for best paper on Implications of Research for Educational Practice from ASTE, and the 2006 Outstanding Early Career Scholar Award from SSMA. Her research focuses on STEM education policy implementation, effective science teaching, and integrated STEM approaches.

Dr. Janet B. Walton is the research assistant professor and the assistant director of evaluation for AEOP at Purdue University's College of Education. Formerly the STEM workforce program manager for Virginia's Region 2000 and founding director of the Future Focus Foundation, a nonprofit organization dedicated to enhancing the quality of STEM education in the region, she merges her economic development and education backgrounds to develop K–12 curricular materials that integrate real-life issues with sound cross-curricular content. Her research focuses on collaboration between schools and community stakeholders for STEM education and problem- and project-based learning pedagogies. With this research agenda, she works to forge productive relationships between K–12 schools and local business and community stakeholders to bring contextual STEM experiences into the classroom and provide students and educators with innovative resources and curricular materials.

Dr. Erin Peters-Burton is the Donna R. and David E. Sterling endowed professor in science education at George Mason University in Fairfax, Virginia. She uses her experiences from 15 years as an engineer and secondary science, engineering, and mathematics teacher to develop research projects that directly inform classroom practice in science and engineering. Her research agenda is based on the idea that all students should build self-awareness of how they learn science and engineering. She works to help students see themselves as "science-minded" and help teachers create classrooms that support student skills to develop scientific knowledge. To accomplish this, she pursues research projects that investigate ways that students and teachers can use self-regulated learning theory in science and engineering, as well as how inclusive STEM schools can help students succeed. During her tenure as a secondary teacher, she had a National Board Certification in Early Adolescent Science and was an Albert Einstein Distinguished Educator Fellow for NASA. As a researcher, Dr. Peters-Burton has published over 100 articles, books, book chapters, and curriculum books focused on STEM education and educational psychology. She received the Outstanding Science Teacher Educator of the Year award from ASTE in 2016 and a Teacher of Distinction Award and a Scholarly Achievement Award from George Mason University in 2012, and in 2010 she was named University Science Educator of the Year by the Virginia Association of Science Teachers.

Tamara J. Moore is an associate professor of engineering education in the College of Engineering at Purdue University. Dr. Moore's research focuses on defining STEM integration through the use of engineering as the connection and investigating its power for student learning.

Toni A. Sondergeld is an associate professor of assessment, research, and statistics in the School of Education at Drexel University in Philadelphia. Dr. Sondergeld's research concentrates on assessment and evaluation in education, with a focus on K–12 STEM.

ACKNOWLEDGMENTS

This module was developed as a part of the STEM Road Map project (Carla C. Johnson, principal investigator). The Purdue University College of Education, General Motors, and other sources provided funding for this project.

See *www.routledge.com/products/9781138804234* for more information about *STEM Road Map: A framework for integrated STEM education.*

PART 1

THE STEM ROAD MAP
BACKGROUND, THEORY, AND PRACTICE

OVERVIEW OF THE *STEM ROAD MAP CURRICULUM SERIES*

Carla C. Johnson, Erin Peters-Burton, and Tamara J. Moore

The *STEM Road Map Curriculum Series* was conceptualized and developed by a team of STEM educators from across the United States in response to a growing need to infuse real-world learning contexts, delivered through authentic problem-solving pedagogy, into K–12 classrooms. The curriculum series is grounded in integrated STEM, which focuses on the integration of the STEM disciplines—science, technology, engineering, and mathematics—delivered across content areas, incorporating the Framework for 21st Century Learning along with grade-level-appropriate academic standards.

The curriculum series begins in kindergarten, with a five-week instructional sequence that introduces students to the STEM themes and gives them grade-level-appropriate topics and real-world challenges or problems to solve. The series uses project-based and problem-based learning, presenting students with the problem or challenge during the first lesson, and then teaching them science, social studies, English language arts, mathematics, and other content, as they apply what they learn to the challenge or problem at hand.

Authentic assessment and differentiation are embedded throughout the modules. Each *STEM Road Map Curriculum Series* module has a lead discipline, which may be science, social studies, English language arts, or mathematics. All disciplines are integrated into each module, along with ties to engineering. Another key component is the use of STEM Research Notebooks to allow students to track their own learning progress. The modules are designed with a scaffolded approach, with increasingly complex concepts and skills introduced as students progress through grade levels.

The developers of this work view the curriculum as a resource that is intended to be used either as a whole or in part to meet the needs of districts, schools, and teachers who are implementing an integrated STEM approach. A variety of implementation formats are possible, from using one stand-alone module at a given grade level to using all five modules to provide 25 weeks of instruction. Also, within each grade band (K–2, 3–5, 6–8, 9–12), the modules can be sequenced in various ways to suit specific needs.

STANDARDS-BASED APPROACH

The *STEM Road Map Curriculum Series* is anchored in the *Next Generation Science Standards (NGSS)*, the *Common Core State Standards for Mathematics (CCSS Mathematics)*, the *Common Core State Standards for English Language Arts (CCSS ELA)*, and the Framework for 21st Century Learning. Each module includes a detailed curriculum map that incorporates the associated standards from the particular area correlated to lesson plans. The STEM Road Map has very clear and strong connections to these academic standards, and each of the grade-level topics was derived from the mapping of the standards to ensure alignment among topics, challenges or problems, and the required academic standards for students. Therefore, the curriculum series takes a standards-based approach and is designed to provide authentic contexts for application of required knowledge and skills.

THEMES IN THE *STEM ROAD MAP CURRICULUM SERIES*

The K–12 STEM Road Map is organized around five real-world STEM themes that were generated through an examination of the big ideas and challenges for society included in STEM standards and those that are persistent dilemmas for current and future generations:

- Cause and Effect
- Innovation and Progress
- The Represented World
- Sustainable Systems
- Optimizing the Human Experience

These themes are designed as springboards for launching students into an exploration of real-world learning situated within big ideas. Most important, the five STEM Road Map themes serve as a framework for scaffolding STEM learning across the K–12 continuum.

The themes are distributed across the STEM disciplines so that they represent the big ideas in science (Cause and Effect; Sustainable Systems), technology (Innovation and Progress; Optimizing the Human Experience), engineering (Innovation and Progress; Sustainable Systems; Optimizing the Human Experience), and mathematics (The Represented World), as well as concepts and challenges in social studies and 21st century skills that are also excellent contexts for learning in English language arts. The process of developing themes began with the clustering of the *NGSS* performance expectations and the National Academy of Engineering's grand challenges for engineering, which led to the development of the challenge in each module and connections of the module activities to the *CCSS Mathematics* and *CCSS ELA* standards. We performed these mapping processes with large teams of experts and found that these five themes provided breadth, depth, and coherence to frame a high-quality STEM learning experience from kindergarten through 12th grade.

Cause and Effect

The concept of cause and effect is a powerful and pervasive notion in the STEM fields. It is the foundation of understanding how and why things happen as they do. Humans spend considerable effort and resources trying to understand the causes and effects of natural and designed phenomena to gain better control over events and the environment and to be prepared to react appropriately. Equipped with the knowledge of a specific cause-and-effect relationship, we can lead better lives or contribute to the community by altering the cause, leading to a different effect. For example, if a person recognizes that irresponsible energy consumption leads to global climate change, that person can act to remedy his or her contribution to the situation. Although cause and effect is a core idea in the STEM fields, it can actually be difficult to determine. Students should be capable of understanding not only when evidence points to cause and effect but also when evidence points to relationships but not direct causality. The major goal of education is to foster students to be empowered, analytic thinkers, capable of thinking through complex processes to make important decisions. Understanding causality, as well as when it cannot be determined, will help students become better consumers, global citizens, and community members.

Innovation and Progress

One of the most important factors in determining whether humans will have a positive future is innovation. Innovation is the driving force behind progress, which helps create possibilities that did not exist before. Innovation and progress are creative entities, but in the STEM fields, they are anchored by evidence and logic, and they use established concepts to move the STEM fields forward. In creating something new, students must consider what is already known in the STEM fields and apply this knowledge appropriately. When we innovate, we create value that was not there previously and create new conditions and possibilities for even more innovations. Students should consider how their innovations might affect progress and use their STEM thinking to change current human burdens to benefits. For example, if we develop more efficient cars that use byproducts from another manufacturing industry, such as food processing, then we have used waste productively and reduced the need for the waste to be hauled away, an indirect benefit of the innovation.

The Represented World

When we communicate about the world we live in, how the world works, and how we can meet the needs of humans, sometimes we can use the actual phenomena to explain a concept. Sometimes, however, the concept is too big, too slow, too small, too fast, or too complex for us to explain using the actual phenomena, and we must use a representation or a model to help communicate the important features. We need representations and models such as graphs, tables, mathematical expressions, and diagrams because it makes our thinking visible. For example, when examining geologic time, we cannot actually observe the passage of such large chunks of time, so we create a timeline or a model that uses a proportional scale to visually illustrate how much time has passed for different eras. Another example may be something too complex for students at a particular grade level, such as explaining the *p* subshell orbitals of electrons to fifth graders. Instead, we use the Bohr model, which more closely represents the orbiting of planets and is accessible to fifth graders.

When we create models, they are helpful because they point out the most important features of a phenomenon. We also create representations of the world with mathematical functions, which help us change parameters to suit the situation. Creating representations of a phenomenon engages students because they are able to identify the important features of that phenomenon and communicate them directly. But because models are estimates of a phenomenon, they leave out some of the details, so it is important for students to evaluate their usefulness as well as their shortcomings.

Sustainable Systems

From an engineering perspective, the term *system* refers to the use of "concepts of component need, component interaction, systems interaction, and feedback. The interaction of subcomponents to produce a functional system is a common lens used by all engineering disciplines for understanding, analysis, and design." (Koehler, Bloom, and Binns 2013, p. 8). Systems can be either open (e.g., an ecosystem) or closed (e.g., a car battery). Ideally, a system should be sustainable, able to maintain equilibrium without much energy from outside the structure. Looking at a garden, we see flowers blooming, weeds sprouting, insects buzzing, and various forms of life living within its boundaries. This is an example of an ecosystem, a collection of living organisms that survive together, functioning as a system. The interaction of the organisms within the system and the influences of the environment (e.g., water, sunlight) can maintain the system for a period of time, thus demonstrating its ability to endure. Sustainability is a desirable feature of a system because it allows for existence of the entity in the long term.

In the STEM Road Map project, we identified different standards that we consider to be oriented toward systems that students should know and understand in the K–12 setting. These include ecosystems, the rock cycle, Earth processes (such as erosion, tectonics, ocean currents, weather phenomena), Earth-Sun-Moon cycles, heat transfer, and the interaction among the geosphere, biosphere, hydrosphere, and atmosphere. Students and teachers should understand that we live in a world of systems that are not independent of each other, but rather are intrinsically linked such that a disruption in one part of a system will have reverberating effects on other parts of the system.

Optimizing the Human Experience

Science, technology, engineering, and mathematics as disciplines have the capacity to continuously improve the ways humans live, interact, and find meaning in the world, thus working to optimize the human experience. This idea has two components: being more suited to our environment and being more fully human. For example, the progression of STEM ideas can help humans create solutions to complex problems, such as improving ways to access water sources, designing energy sources with minimal impact on our environment, developing new ways of communication and expression, and building efficient shelters. STEM ideas can also provide access to the secrets and wonders of nature. Learning in STEM requires students to think logically and systematically, which is a way of knowing the world that is markedly different from knowing the world as an artist. When students can employ various ways of knowing and understand when it is appropriate to use a different way of knowing or integrate ways of knowing, they are fully experiencing the best of what it is to be human. The problem-based learning scenarios provided in the STEM Road Map help students develop ways of thinking like STEM professionals as they ask questions and design solutions. They learn to optimize the human experience by innovating improvements in the designed world in which they live.

THE NEED FOR AN INTEGRATED STEM APPROACH

At a basic level, STEM stands for science, technology, engineering, and mathematics. Over the past decade, however, STEM has evolved to have a much broader scope and implications. Now, educators and policy makers refer to STEM as not only a concentrated area for investing in the future of the United States and other nations but also as a domain and mechanism for educational reform.

The good intentions of the recent decade-plus of focus on accountability and increased testing has resulted in significant decreases not only in instructional time for teaching science and social studies but also in the flexibility of teachers to promote authentic, problem solving–focused classroom environments. The shift has had a detrimental impact on student acquisition of vitally important skills, which many refer to as 21st century skills, and often the ability of students to "think." Further, schooling has become increasingly siloed into compartments of mathematics, science, English language arts, and social studies, lacking any of the connections that are overwhelmingly present in

the real world around children. Students have experienced school as content provided in boxes that must be memorized, devoid of any real-world context, and often have little understanding of why they are learning these things.

STEM-focused projects, curriculum, activities, and schools have emerged as a means to address these challenges. However, most of these efforts have continued to focus on the individual STEM disciplines (predominantly science and engineering) through more STEM classes and after-school programs in a "STEM enhanced" approach (Breiner et al. 2012). But in traditional and STEM enhanced approaches, there is little to no focus on other disciplines that are integral to the context of STEM in the real world. Integrated STEM education, on the other hand, infuses the learning of important STEM content and concepts with a much-needed emphasis on 21st century skills and a problem- and project-based pedagogy that more closely mirrors the real-world setting for society's challenges. It incorporates social studies, English language arts, and the arts as pivotal and necessary (Johnson 2013; Rennie, Venville, and Wallace 2012; Roehrig et al. 2012).

FRAMEWORK FOR STEM INTEGRATION IN THE CLASSROOM

The *STEM Road Map Curriculum Series* is grounded in the Framework for STEM Integration in the Classroom as conceptualized by Moore, Guzey, and Brown (2014) and Moore et al. (2014). The framework has six elements, described in the context of how they are used in the *STEM Road Map Curriculum Series* as follows:

1. The STEM Road Map contexts are meaningful to students and provide motivation to engage with the content. Together, these allow students to have different ways to enter into the challenge.

2. The STEM Road Map modules include engineering design that allows students to design technologies (i.e., products that are part of the designed world) for a compelling purpose.

3. The STEM Road Map modules provide students with the opportunities to learn from failure and redesign based on the lessons learned.

4. The STEM Road Map modules include standards-based disciplinary content as the learning objectives.

5. The STEM Road Map modules include student-centered pedagogies that allow students to grapple with the content, tie their ideas to the context, and learn to think for themselves as they deepen their conceptual knowledge.

6. The STEM Road Map modules emphasize 21st century skills and, in particular, highlight communication and teamwork.

All of the STEM Road Map modules incorporate these six elements; however, the level of emphasis on each of these elements varies based on the challenge or problem in each module.

THE NEED FOR THE *STEM ROAD MAP CURRICULUM SERIES*

As focus is increasing on integrated STEM, and additional schools and programs decide to move their curriculum and instruction in this direction, there is a need for quality, research-based curriculum designed with integrated STEM at the core. Several good resources are available to help teachers infuse engineering or more STEM enhanced approaches, but no curriculum exists that spans K–12 with an integrated STEM focus. The next chapter provides detailed information about the specific pedagogy, instructional strategies, and learning theory on which the *STEM Road Map Curriculum Series* is grounded.

REFERENCES

Breiner, J., M. Harkness, C. C. Johnson, and C. Koehler. 2012. What is STEM? A discussion about conceptions of STEM in education and partnerships. *School Science and Mathematics* 112 (1): 3–11.

Johnson, C. C. 2013. Conceptualizing integrated STEM education: Editorial. *School Science and Mathematics* 113 (8): 367–368.

Koehler, C. M., M. A. Bloom, and I. C. Binns. 2013. Lights, camera, action: Developing a methodology to document mainstream films' portrayal of nature of science and scientific inquiry. *Electronic Journal of Science Education* 17 (2).

Moore, T. J., S. S. Guzey, and A. Brown. 2014. Greenhouse design to increase habitable land: An engineering unit. *Science Scope* 51–57.

Moore, T. J., M. S. Stohlmann, H.-H. Wang, K. M. Tank, A. W. Glancy, and G. H. Roehrig. 2014. Implementation and integration of engineering in K–12 STEM education. In *Engineering in pre-college settings: Synthesizing research, policy, and practices,* ed. S. Purzer, J. Strobel, and M. Cardella, 35–60. West Lafayette, IN: Purdue Press.

Rennie, L., G. Venville, and J. Wallace. 2012. *Integrating science, technology, engineering, and mathematics: Issues, reflections, and ways forward.* New York: Routledge.

Roehrig, G. H., T. J. Moore, H. H. Wang, and M. S. Park. 2012. Is adding the *E* enough? Investigating the impact of K–12 engineering standards on the implementation of STEM integration. *School Science and Mathematics* 112 (1): 31–44.

STRATEGIES USED IN THE
STEM ROAD MAP CURRICULUM SERIES

Erin Peters-Burton, Carla C. Johnson, Toni A. Sondergeld, and Tamara J. Moore

The *STEM Road Map Curriculum Series* uses what has been identified through research as best-practice pedagogy, including embedded formative assessment strategies throughout each module. This chapter briefly describes the key strategies that are employed in the series.

PROJECT- AND PROBLEM-BASED LEARNING

Each module in the *STEM Road Map Curriculum Series* uses either project-based learning or problem-based learning to drive the instruction. Project-based learning begins with a driving question to guide student teams in addressing a contextualized local or community problem or issue. The outcome of project-based instruction is a product that is conceptualized, designed, and tested through a series of scaffolded learning experiences (Blumenfeld et al. 1991; Krajcik and Blumenfeld 2006). Problem-based learning is often grounded in a fictitious scenario, challenge, or problem (Barell 2006; Lambros 2004). On the first day of instruction within the unit, student teams are provided with the context of the problem. Teams work through a series of activities and use open-ended research to develop their potential solution to the problem or challenge, which need not be a tangible product (Johnson 2003).

ENGINEERING DESIGN PROCESS

The *STEM Road Map Curriculum Series* uses engineering design as a way to facilitate integrated STEM within the modules. The engineering design process (EDP) is depicted in Figure 2.1 (p. 10). It highlights two major aspects of engineering design—problem scoping and solution generation—and six specific components of working toward a design: define the problem, learn about the problem, plan a solution, try the solution, test the solution, decide whether the solution is good enough. It also shows that communication

Figure 2.1. Engineering Design Process

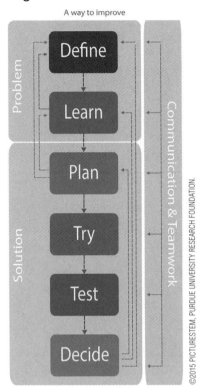

©2015 PICTURESTEM, PURDUE UNIVERSITY RESEARCH FOUNDATION.

and teamwork are involved throughout the entire process. As the arrows in the figure indicate, the order in which the components of engineering design are addressed depends on what becomes needed as designers progress through the EDP. Designers must communicate and work in teams throughout the process. The EDP is iterative, meaning that components of the process can be repeated as needed until the design is good enough to present to the client as a potential solution to the problem.

Problem scoping is the process of gathering and analyzing information to deeply understand the engineering design problem. It includes defining the problem and learning about the problem. Defining the problem includes identifying the problem, the client, and the end user of the design. The client is the person (or people) who hired the designers to do the work, and the end user is the person (or people) who will use the final design. The designers must also identify the criteria and the constraints of the problem. The criteria are the things the client wants from the solution, and the constraints are the things that limit the possible solutions. The designers must spend significant time learning about the problem, which can include activities such as the following:

- Reading informational texts and researching about relevant concepts or contexts

- Identifying and learning about needed mathematical and scientific skills, knowledge, and tools

- Learning about things done previously to solve similar problems

- Experimenting with possible materials that could be used in the design

Problem scoping also allows designers to consider how to measure the success of the design in addressing specific criteria and staying within the constraints over multiple iterations of solution generation.

Solution generation includes planning a solution, trying the solution, testing the solution, and deciding whether the solution is good enough. Planning the solution includes generating many design ideas that both address the criteria and meet the constraints. Here the designers must consider what was learned about the problem during problem scoping. Design plans include clear communication of design ideas through media such as notebooks, blueprints, schematics, or storyboards. They also include details about the design, such as measurements, materials, colors, costs of materials, instructions for how things fit together, and sets of directions. Making the decision about which design idea to move forward involves considering the trade-offs of each design idea.

Once a clear design plan is in place, the designers must try the solution. Trying the solution includes developing a prototype (a testable model) based on the plan generated. The prototype might be something physical or a process to accomplish a goal. This component of design requires that the designers consider the risk involved in implementing the design. The prototype developed must be tested. Testing the solution includes conducting fair tests that verify whether the plan is a solution that is good enough to meet the client and end user needs and wants. Data need to be collected about the results of the tests of the prototype, and these data should be used to make evidence-based decisions regarding the design choices made in the plan. Here, the designers must again consider the criteria and constraints for the problem.

Using the data gathered from the testing, the designers must decide whether the solution is good enough to meet the client and end user needs and wants by assessment based on the criteria and constraints. Here, the designers must justify or reject design decisions based on the background research gathered while learning about the problem and on the evidence gathered during the testing of the solution. The designers must now decide whether to present the current solution to the client as a possibility or to do more iterations of design on the solution. If they decide that improvements need to be made to the solution, the designers must decide if there is more that needs to be understood about the problem, client, or end user; if another design idea should be tried; or if more planning needs to be conducted on the same design. One way or another, more work needs to be done.

Throughout the process of designing a solution to meet a client's needs and wants, designers work in teams and must communicate to each other, the client, and likely the end user. Teamwork is important in engineering design because multiple perspectives and differing skills and knowledge are valuable when working to solve problems. Communication is key to the success of the designed solution. Designers must communicate their ideas clearly using many different representations, such as text in an engineering notebook, diagrams, flowcharts, technical briefs, or memos to the client.

LEARNING CYCLE

The same format for the learning cycle is used in all grade levels throughout the STEM Road Map, so that students engage in a variety of activities to learn about phenomena in the modules thoroughly and have consistent experiences in the problem- and project-based learning modules. Expectations for learning by younger students are not as high as for older students, but the format of the progression of learning is the same. Students who have learned with curriculum from the STEM Road Map in early grades know

what to expect in later grades. The learning cycle consists of five parts—Introductory Activity/Engagement, Activity/Exploration, Explanation, Elaboration/Application of Knowledge, and Evaluation/Assessment—and is based on the empirically tested 5E model from BSCS (Bybee et al. 2006).

In the Introductory Activity/Engagement phase, teachers introduce the module challenge and use a unique approach designed to pique students' curiosity. This phase gets students to start thinking about what they already know about the topic and begin wondering about key ideas. The Introductory Activity/Engagement phase positions students to be confident about what they are about to learn, because they have prior knowledge, and clues them into what they don't yet know.

In the Activity/Exploration phase, the teacher sets up activities in which students experience a deeper look at the topics that were introduced earlier. Students engage in the activities and generate new questions or consider possibilities using preliminary investigations. Students work independently, in small groups, and in whole-group settings to conduct investigations, resulting in common experiences about the topic and skills involved in the real-world activities. Teachers can assess students' development of concepts and skills based on the common experiences during this phase.

During the Explanation phase, teachers direct students' attention to concepts they need to understand and skills they need to possess to accomplish the challenge. Students participate in activities to demonstrate their knowledge and skills to this point, and teachers can pinpoint gaps in student knowledge during this phase.

In the Elaboration/Application of Knowledge phase, teachers present students with activities that engage in higher-order thinking to create depth and breadth of student knowledge, while connecting ideas across topics within and across STEM. Students apply what they have learned thus far in the module to a new context or elaborate on what they have learned about the topic to a deeper level of detail.

In the last phase, Assessment, teachers give students summative feedback on their knowledge and skills as demonstrated through the challenge. This is not the only point of assessment (as discussed in the section on Embedded Formative Assessments), but it is an assessment of the culmination of the knowledge and skills for the module. Students demonstrate their cognitive growth at this point and reflect on how far they have come since the beginning of the module. The challenges are designed to be multidimensional in the ways students must collaborate and communicate their new knowledge.

STEM RESEARCH NOTEBOOK

One of the main components of the *STEM Road Map Curriculum Series* is the STEM Research Notebook, a place for students to capture their ideas, questions, observations, reflections, evidence of progress, and other items associated with their daily work. At the beginning of each module, the teacher walks students through the setup of the STEM

Research Notebook, which could be a three-ring binder, composition book, or spiral notebook. You may wish to have students create divided sections so that they can easily access work from various disciplines during the module. Electronic notebooks kept on student devices are also acceptable and encouraged. Students will develop their own table of contents and create chapters in the notebook for each module.

Each lesson in the *STEM Road Map Curriculum Series* includes one or more prompts that are designed for inclusion in the STEM Research Notebook and appear as questions or statements that the teacher assigns to students. These prompts require students to apply what they have learned across the lesson to solve the big problem or challenge for that module. Each lesson is designed to meaningfully refer students to the larger problem or challenge they have been assigned to solve with their teams. The STEM Research Notebook is designed to be a key formative assessment tool, as students' daily entries provide evidence of what they are learning. The notebook can be used as a mechanism for dialogue between the teacher and students, as well as for peer and self-evaluation.

The use of the STEM Research Notebook is designed to scaffold student notebooking skills across the grade bands in the *STEM Road Map Curriculum Series*. In the early grades, children learn how to organize their daily work in the notebook as a way to collect their products for future reference. In elementary school, students structure their notebooks to integrate background research along with their daily work and lesson prompts. In the upper grades (middle and high school), students expand their use of research and data gathering through team discussions to more closely mirror the work of STEM experts in the real world.

THE ROLE OF ASSESSMENT IN THE *STEM ROAD MAP CURRICULUM SERIES*

Starting in the middle years and continuing into secondary education, the word *assessment* typically brings grades to mind. These grades may take the form of a letter or a percentage, but they typically are used as a representation of a student's content mastery. If well thought out and implemented, however, classroom assessment can offer teachers, parents, and students valuable information about student learning and misconceptions that does not necessarily come in the form of a grade (Popham 2013).

The *STEM Road Map Curriculum Series* provides a set of assessments for each module. Teachers are encouraged to use assessment information for more than just assigning grades to students. Instead, assessments of activities requiring students to actively engage in their learning, such as student journaling in STEM Research Notebooks, collaborative presentations, and constructing graphic organizers, should be used to move student learning forward. Whereas other curriculum with assessments may include objective-type (multiple-choice or matching) tests, quizzes, or worksheets, we have intentionally avoided these forms of assessments to better align assessment strategies with teacher instruction and student learning techniques. Since the focus of this book is on project- or problem-based STEM curriculum and instruction that focuses on higher-level thinking skills, appropriate and authentic performance assessments were developed to elicit the most reliable and valid indication of growth in student abilities (Brookhart and Nitko 2008).

Comprehensive Assessment System

Assessment throughout all STEM Road Map curriculum modules acts as a comprehensive system in which formative and summative assessments work together to provide teachers with high-quality information on student learning. Formative assessment occurs when the teacher finds out formally or informally what a student knows about a smaller, defined concept or skill and provides timely feedback to the student about his or her level of proficiency. Summative assessments occur when students have performed all activities in the module and are given a cumulative performance evaluation in which they demonstrate their growth in learning.

A comprehensive assessment system can be thought of as akin to a sporting event. Formative assessments are the practices: It is important to accomplish them consistently, they provide feedback to help students improve their learning, and making mistakes can be worthwhile if students are given an opportunity to learn from them. Summative assessments are the competitions: Students need to be prepared to perform at the best of their ability. Without multiple opportunities to practice skills along the way through formative assessments, students will not have the best chance of demonstrating growth in abilities through summative assessments (Black and Wiliam 1998).

Embedded Formative Assessments

Formative assessments in this module serve two main purposes: to provide feedback to students about their learning and to provide important information for the teacher to inform immediate instructional needs. Providing feedback to students is particularly important when conducting problem- or project-based learning because students take on much of the responsibility for learning, and teachers must facilitate student learning in an informed way. For example, if students are required to conduct research for the Activity/Exploration phase but are not familiar with what constitutes a reliable resource, they may develop misconceptions based on poor information. When a teacher monitors this learning through formative assessments and provides specific feedback related to the instructional goals, students are less likely to develop incomplete or incorrect conceptions in their independent investigations. By using formative assessment to detect problems in student learning and then acting on this information, teachers help move student learning forward through these teachable moments.

Formative assessments come in a variety of formats. They can be informal, such as asking students probing questions related to student knowledge or tasks or simply observing students engaged in an activity to gather information about student skills. Formative assessments can also be formal, such as a written quiz or a laboratory practical. Regardless of the type, three key steps must be completed when using formative assessments (Sondergeld, Bell, and Leusner 2010). First, the assessment is delivered to students so that teachers can collect data. Next, teachers analyze the data (student responses) to determine student strengths and areas that need additional support. Finally, teachers use the results from information collected to modify lessons and create learning environments that reinforce weak points in student learning. If student learning information is not used to modify instruction, the assessment cannot be considered formative in nature.

Formative assessments can be about content, science process skills, or even learning skills. When a formative assessment focuses on content, it assesses student knowledge about the disciplinary core ideas from the *Next Generation Science Standards* (*NGSS*) or content objectives from *Common Core State Standards for Mathematics* (*CCSS Mathematics*) or *Common Core State Standards for English Language Arts* (*CCSS ELA*). Content-focused formative assessments ask students questions about declarative knowledge regarding the concepts they have been learning. Process skills formative assessments examine the extent to which a student can perform science and engineering practices from the *NGSS* or process objectives from *CCSS Mathematics* or *CCSS ELA*, such as constructing an argument. Learning skills can also be assessed formatively by asking students to reflect on the ways they learn best during a module and identify ways they could have learned more.

Assessment Maps

Assessment maps or blueprints can be used to ensure alignment between classroom instruction and assessment. If what students are learning in the classroom is not the same as the content on which they are assessed, the resultant judgment made on student learning will be invalid (Brookhart and Nitko 2008). Therefore, the issue of instruction and assessment alignment is critical. The assessment map for this book (found in Chapter 3) indicates by lesson whether the assessment should be completed as a group or on an individual basis, identifies the assessment as formative or summative in nature, and aligns the assessment with its corresponding learning objectives.

Note that the module includes far more formative assessments than summative assessments. This is done intentionally to provide students with multiple opportunities to practice their learning of new skills before completing a summative assessment. Note also that formative assessments are used to collect information on only one or two learning objectives at a time so that potential relearning or instructional modifications can focus on smaller and more manageable chunks of information. Conversely, summative assessments

in the module cover many more learning objectives, as they are traditionally used as final markers of student learning. This is not to say that information collected from summative assessments cannot or should not be used formatively. If teachers find that gaps in student learning persist after a summative assessment is completed, it is important to revisit these existing misconceptions or areas of weakness before moving on (Black et al. 2003).

SELF-REGULATED LEARNING THEORY IN THE STEM ROAD MAP MODULES

Many learning theories are compatible with the STEM Road Map modules, such as constructivism, situated cognition, and meaningful learning. However, we feel that the self-regulated learning theory (SRL) aligns most appropriately (Zimmerman 2000). SRL requires students to understand that thinking needs to be motivated and managed (Ritchhart, Church, and Morrison 2011). The STEM Road Map modules are student centered and are designed to provide students with choices, concrete hands-on experiences, and opportunities to see and make connections, especially across subjects (Eliason and Jenkins 2012; NAEYC 2016). Additionally, SRL is compatible with the modules because it fosters a learning environment that supports students' motivation, enables students to become aware of their own learning strategies, and requires reflection on learning while experiencing the module (Peters and Kitsantas 2010).

The theory behind SRL (see Figure 2.2) explains the different processes that students engage in before, during, and after a learning task. Because SRL is a cyclical learning process, the accomplishment of one cycle develops strategies for the next learning cycle. This cyclic way of learning aligns with the various sections in the STEM Road Map lesson plans on Introductory Activity/Engagement, Activity/Exploration, Explanation, Elaboration/Application of Knowledge, and Evaluation/Assessment. Since the students engaged in a module take on much

Figure 2.2. SRL Theory

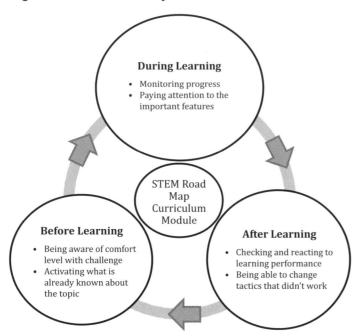

Source: Adapted from Zimmerman 2000.

of the responsibility for learning, this theory also provides guidance for teachers to keep students on the right track.

The remainder of this section explains how SRL theory is embedded within the five sections of each module and points out ways to support students in becoming independent learners of STEM while productively functioning in collaborative teams.

Before Learning: Setting the Stage

Before attempting a learning task such as the STEM Road Map modules, teachers should develop an understanding of their students' level of comfort with the process of accomplishing the learning and determine what they already know about the topic. When students are comfortable with attempting a learning task, they tend to take more risks in learning and as a result achieve deeper learning (Bandura 1986).

The STEM Road Map curriculum modules are designed to foster excitement from the very beginning. Each module has an Introductory Activity/Engagement section that introduces the overall topic from a unique and exciting perspective, engaging the students to learn more so that they can accomplish the challenge. The Introductory Activity also has a design component that helps teachers assess what students already know about the topic of the module. In addition to the deliberate designs in the lesson plans to support SRL, teachers can support a high level of student comfort with the learning challenge by finding out if students have ever accomplished the same kind of task and, if so, asking them to share what worked well for them.

During Learning: Staying the Course

Some students fear inquiry learning because they aren't sure what to do to be successful (Peters 2010). However, the STEM Road Map curriculum modules are embedded with tools to help students pay attention to knowledge and skills that are important for the learning task and to check student understanding along the way. One of the most important processes for learning is the ability for learners to monitor their own progress while performing a learning task (Peters 2012). The modules allow students to monitor their progress with tools such as the STEM Research Notebooks, in which they record what they know and can check whether they have acquired a complete set of knowledge and skills. The STEM Road Map modules support inquiry strategies that include previewing, questioning, predicting, clarifying, observing, discussing, and journaling (Morrison and Milner 2014). Through the use of technology throughout the modules, inquiry is supported by providing students access to resources and data while enabling them to process information, report the findings, collaborate, and develop 21st century skills.

It is important for teachers to encourage students to have an open mind about alternative solutions and procedures (Milner and Sondergeld 2015) when working through the STEM Road Map curriculum modules. Novice learners can have difficulty knowing

what to pay attention to and tend to treat each possible avenue for information as equal (Benner 1984). Teachers are the mentors in a classroom and can point out ways for students to approach learning during the Activity/Exploration, Explanation, and Elaboration/Application of Knowledge portions of the lesson plans to ensure that students pay attention to the important concepts and skills throughout the module. For example, if a student is to demonstrate conceptual awareness of motion when working on roller coaster research, but the student has misconceptions about motion, the teacher can step in and redirect student learning.

After Learning: Knowing What Works

The classroom is a busy place, and it may often seem that there is no time for self-reflection on learning. Although skipping this reflective process may save time in the short term, it reduces the ability to take into account things that worked well and things that didn't so that teaching the module may be improved next time. In the long run, SRL skills are critical for students to become independent learners who can adapt to new situations. By investing the time it takes to teach students SRL skills, teachers can save time later, because students will be able to apply methods and approaches for learning that they have found effective to new situations. In the Evaluation/Assessment portion of the STEM Road Map curriculum modules, as well as in the formative assessments throughout the modules, two processes in the after-learning phase are supported: evaluating one's own performance and accounting for ways to adapt tactics that didn't work well. Students have many opportunities to self-assess in formative assessments, both in groups and individually, using the rubrics provided in the modules.

The designs of the *NGSS* and *CCSS* allow for students to learn in diverse ways, and the STEM Road Map curriculum modules emphasize that students can use a variety of tactics to complete the learning process. For example, students can use STEM Research Notebooks to record what they have learned during the various research activities. Notebook entries might include putting objectives in students' own words, compiling their prior learning on the topic, documenting new learning, providing proof of what they learned, and reflecting on what they felt successful doing and what they felt they still needed to work on. Perhaps students didn't realize that they were supposed to connect what they already knew with what they learned. They could record this and would be prepared in the next learning task to begin connecting prior learning with new learning.

SAFETY IN STEM

Student safety is a primary consideration in all subjects but is an area of particular concern in science, where students may interact with unfamiliar tools and materials that may pose additional safety risks. It is important to implement safety practices within the context of STEM investigations, whether in a classroom laboratory or in the field. When you keep safety in mind as a teacher, you avoid many potential issues with the lesson while also protecting your students.

STEM safety practices encompass things considered in the typical science classroom. Ensure that students are familiar with basic safety considerations, such as wearing protective equipment (e.g., safety glasses or goggles and latex-free gloves) and taking care with sharp objects, and know emergency exit procedures. Teachers should learn beforehand the locations of the safety eyewash, fume hood, fire extinguishers, and emergency shut-off switch in the classroom and how to use them. Also be aware of any school or district safety policies that are in place and apply those that align with the work being conducted in the lesson. It is important to review all safety procedures annually.

STEM investigations should always be supervised. Each lesson in the modules includes teacher guidelines for applicable safety procedures that should be followed. Before each investigation, teachers should go over these safety procedures with the student teams. Some STEM focus areas such as engineering require that students can demonstrate how to properly use equipment in the maker space before the teacher allows them to proceed with the lesson.

Information about classroom science safety, including a safety checklist for science classrooms, general lab safety recommendations, and links to other science safety resources, is available at the Council of State Science Supervisors (CSSS) website at *www.csss-science.org/safety.shtml.* The National Science Teachers Association (NSTA) provides a list of science rules and regulations, including standard operating procedures for lab safety, and a safety acknowledgement form for students and parents or guardians to sign. You can access this forum at *http://static.nsta.org/pdfs/SafetyInTheScienceClassroomLabAndField.pdf.* In addition, NSTA's Safety in the Science Classroom web page (*www.nsta.org/safety*) has numerous links to safety resources, including papers written by the NSTA Safety Advisory Board.

Disclaimer: The safety precautions for each activity are based on use of the recommended materials and instructions, legal safety standards, and better professional practices. Using alternative materials or procedures for these activities may jeopardize the level of safety and therefore is at the user's own risk. Further information regarding safety procedures can be found in other NSTA publications, such as the guide "Safety in the Science Classroom, Laboratory, or Field" (*http://static.nsta.org/pdfs/SafetyInTheScience Classroom.pdf*).

REFERENCES

Bandura, A. 1986. *Social foundations of thought and action: A social cognitive theory.* Englewood Cliffs, NJ: Prentice-Hall.

Barell, J. 2006. *Problem-based learning: An inquiry approach.* Thousand Oaks, CA: Corwin Press.

Benner, P. 1984. *From novice to expert: Excellence and power in clinical nursing practice.* Menlo Park, CA: Addison-Wesley Publishing Company.

Black, P., C. Harrison, C. Lee, B. Marshall, and D. Wiliam. 2003. *Assessment for learning: Putting it into practice.* Berkshire, UK: Open University Press.

Black, P., and D. Wiliam. 1998. Inside the black box: Raising standards through classroom assessment. *Phi Delta Kappan* 80 (2): 139–148.

Blumenfeld, P., E. Soloway, R. Marx, J. Krajcik, M. Guzdial, and A. Palincsar. 1991. Motivating project-based learning: Sustaining the doing, supporting learning. *Educational Psychologist* 26 (3): 369–398.

Brookhart, S. M., and A.J. Nitko. 2008. *Assessment and grading in classrooms.* Upper Saddle River, NJ: Pearson.

Bybee, R., J. Taylor, A. Gardner, P. Van Scotter, J. Carlson, A. Westbrook, and N. Landes. 2006. *The BSCS 5E instructional model: Origins and effectiveness. http://science.education.nih.gov/houseofreps. nsf/b82d55fa138783c2852572c9004f5566/$FILE/Appendix?D.pdf.*

Eliason, C. F., and L. T. Jenkins. 2012. *A practical guide to early childhood curriculum.* 9th ed. New York: Merrill.

Johnson, C. 2003. Bioterrorism is real-world science: Inquiry-based simulation mirrors real life. *Science Scope* 27 (3): 19–23.

Krajcik, J., and P. Blumenfeld. 2006. Project-based learning. In *The Cambridge handbook of the learning sciences,* ed. R. Keith Sawyer, 317–334. New York: Cambridge University Press.

Lambros, A. 2004. *Problem-based learning in middle and high school classrooms: A teacher's guide to implementation.* Thousand Oaks, CA: Corwin Press.

Milner, A. R., and T. Sondergeld. 2015. Gifted urban middle school students: The inquiry continuum and the nature of science. *National Journal of Urban Education and Practice* 8 (3): 442–461.

Morrison, V., and A. R. Milner. 2014. Literacy in support of science: A closer look at cross-curricular instructional practice. *Michigan Reading Journal* 46 (2): 42–56.

National Association for the Education of Young Children (NAEYC). 2016. Developmentally appropriate practice position statements. *www.naeyc.org/positionstatements/dap.*

Peters, E. E. 2010. Shifting to a student-centered science classroom: An exploration of teacher and student changes in perceptions and practices. *Journal of Science Teacher Education* 21 (3): 329–349.

Peters, E. E. 2012. Developing content knowledge in students through explicit teaching of the nature of science: Influences of goal setting and self-monitoring. *Science and Education* 21 (6): 881–898.

Peters, E. E., and A. Kitsantas. 2010. The effect of nature of science metacognitive prompts on science students' content and nature of science knowledge, metacognition, and self-regulatory efficacy. *School Science and Mathematics* 110: 382–396.

Popham, W. J. 2013. *Classroom assessment: What teachers need to know.* 7th ed. Upper Saddle River, NJ: Pearson.

Ritchhart, R., M. Church, and K. Morrison. 2011. *Making thinking visible: How to promote engagement, understanding, and independence for all learners.* San Francisco, CA: Jossey-Bass.

Sondergeld, T. A., C. A. Bell, and D. M. Leusner. 2010. Understanding how teachers engage in formative assessment. *Teaching and Learning* 24 (2): 72–86.

Zimmerman, B. J. 2000. Attaining self-regulation: A social-cognitive perspective. In *Handbook of self-regulation,* ed. M. Boekaerts, P. Pintrich, and M. Zeidner, 13–39. San Diego: Academic Press.

PART 2

AMUSEMENT PARK OF THE FUTURE

STEM ROAD MAP MODULE

AMUSEMENT PARK OF THE FUTURE MODULE OVERVIEW

Erin Peters-Burton, Janet B. Walton, and Carla C. Johnson

THEME: Innovation and Progress

LEAD DISCIPLINE: Science

MODULE SUMMARY

We use buildings every day, but often take for granted how complex these structures are, and this unit gives students an inside look at the technologies and science necessary to understand these outstanding feats of engineering. In this module, students will examine micro and macro properties of construction materials, particularly those of high-rise buildings. For each subject area, the unit is split into three sections. During the first section, students will learn how high-rises are constructed, the influence these high-rises had on society, and how to communicate complex ideas clearly. In the second section, students will look at the factors for the collapse of the World Trade Center Twin Towers in New York, focusing on how engineers use failure to learn more about the designed world. In the last section, students will examine innovations in construction to propose new ways to construct high-rises (summary adapted from Peters-Burton et al. 2015).

ESTABLISHED GOALS AND OBJECTIVES

At the conclusion of this module, students will be able to do the following:

- Understand the big ideas around energy transfer, including potential and kinetic energy transfer.

- Use their measurement skills to find ratios and rates to describe their prototype amusements and graph the results of their investigations.

- Relate what they have learned about the history of amusements to understand why people seek thrills in their leisure time.

- Practice their English language arts (ELA) skills by understanding technical texts, creating multimedia communication products, and creating arguments for the claims they make based on the evidence of the investigations.

- See how the subjects they study not only provide information about the world around them but also work together to create a more comprehensive understanding of phenomena.

CHALLENGE OR PROBLEM FOR STUDENTS TO SOLVE: AMUSEMENT PARK OF THE FUTURE DESIGN CHALLENGE

Student teams are challenged to each produce a prototype of an amusement park. They begin by conducting research on the advances in amusement parks, starting with 19th-century amusement rides. Students also research the role of amusement parks in society and synthesize their research to inform the creation of their prototypes. They use this research to present their ideas for parks using today's technology, including rides and dart- or ball-throwing games. Students create blueprints of their models, build and test small-scale prototypes, and develop cost-benefit analyses for building and maintaining their parks, including impact studies for the local communities in which the parks will be located. Students also design marketing plans and infomercials with scripts and demonstrations to promote their amusement parks.

Driving Question: How can we use what we know about the development of amusements, the ways people experience thrills, and the laws of physics to propose new amusements that are both safe and extreme?

STEM RESEARCH NOTEBOOK

Each student should maintain a STEM Research Notebook, which will serve as a place for students to organize their work throughout this module (see p. 25 for more general discussion on setup and use of this notebook). All written work in the module should be included in the notebook, including records of students' thoughts and ideas, fictional accounts based on the concepts in the module, and records of student progress through the engineering design process. The notebooks may be maintained across subject areas, giving students the opportunity to see that although their classes may be separated during the school day, the knowledge they gain is connected.

Emphasize to students the importance of organizing all information in a Research Notebook. Explain to them that scientists and other researchers maintain detailed Research Notebooks in their work. These notebooks, which are crucial to researchers' work because they contain critical information and track the researchers' progress, are often considered legal documents for scientists who are pursuing patents or wish to provide proof of their discovery process.

STEM RESEARCH NOTEBOOK GUIDELINES

STEM professionals record their ideas, inventions, experiments, questions, observations, and other work details in notebooks so that they can use these notebooks to help them think about their projects and the problems they are trying to solve. You will each keep a STEM Research Notebook during this module that is like the notebooks that STEM professionals use. In this notebook, you will include all your work and notes about ideas you have. The notebook will help you connect your daily work with the big problem or challenge you are working to solve.

It is important that you organize your notebook entries under the following headings:

1. **Chapter Topic or Title of Problem or Challenge:** You will start a new chapter in your STEM Research Notebook for each new module. This heading is the topic or title of the big problem or challenge that your team is working to solve in this module.

2. **Date and Topic of Lesson Activity for the Day:** Each day, you will begin your daily entry by writing the date and the day's lesson topic at the top of a new page. Write the page number both on the page and in the table of contents.

3. **Information Gathered From Research:** This is information you find from outside resources such as websites or books.

4. **Information Gained From Class or Discussions With Team Members:** This information includes any notes you take in class and notes about things your team discusses. You can include drawings of your ideas here, too.

5. **New Data Collected From Investigations:** This includes data gathered from experiments, investigations, and activities in class.

6. **Documents:** These are handouts and other resources you may receive in class that will help you solve your big problem or challenge. Paste or staple these documents in your STEM Research Notebook for safekeeping and easy access later.

7. **Personal Reflections:** Here, you record your own thoughts and ideas on what you are learning.

8. **Lesson Prompts:** These are questions or statements that your teacher assigns you within each lesson to help you solve your big problem or challenge. You will respond to the prompts in your notebook.

9. **Other Items:** This section includes any other items your teacher gives you or other ideas or questions you may have.

MODULE LAUNCH

To launch the module, have students investigate close-up photos of people riding amusements and then discuss why people seek thrill rides. The main goal of the launch is to connect the idea that people seek emotion-based thrills with the physics and engineering of amusement rides and games and to convey the message that physics and engineering are an integral part of generating these emotions. Next, students watch an artistically derived video of extreme amusements. Although the video and accompanying website appear to be a research project on amusements, the video has actually been manipulated to look like footage of a real park. Nevertheless, the extreme nature of the rides in the video and the ways the "scientists" talk about the rides will engage the interest of students and clearly illustrate how the rides use physical characteristics to evoke thrills.

PREREQUISITE SKILLS FOR THE MODULE

Students enter this module with a wide range of preexisting skills, information, and knowledge. Table 3.1 provides an overview of prerequisite skills and knowledge that students are expected to apply in this module, along with examples of how they apply this knowledge throughout the module. Differentiation strategies are also provided for students who may need additional support in acquiring or applying this knowledge.

Table 3.1. Prerequisite Key Knowledge and Examples of Applications and Differentiation Strategies

Prerequisite Key Knowledge	Application of Knowledge by Students	Differentiation for Students Needing Additional Knowledge
• Use electronic and print media to find new information from reliable sources. • Summarize information gathered from several sources. • Measure linear distance and time in metric units. • Assess and address safety issues relative to amusements. • Use simple arithmetic operations (adding, subtracting, multiplying, dividing).	• Create a timeline for a variety of amusements. • Create a report featuring psychological factors of thrill seeking at amusement parks. • Use the internet find the tallest and the fastest amusement rides in the world, as well as the one with the most loops. • Research safety features of amusements. • Apply measurement and arithmetic operations to build an amusement park prototype.	• Provide a list of reliable sources for students to use. • Highlight key information from reliable resources for students to synthesize. • Review measurement skills and provide opportunities for practice throughout the module. • Review arithmetic operations and provide opportunities to apply operations to real-life situations.

POTENTIAL STEM MISCONCEPTIONS

Students enter the classroom with a wide variety of prior knowledge and ideas, so it is important to be alert to misconceptions, or inappropriate understandings of foundational knowledge. These misconceptions can be classified as one of several types: "preconceived notions," opinions based on popular beliefs or understandings; "nonscientific beliefs," knowledge students have gained about science from sources outside the scientific community; "conceptual misunderstandings," incorrect conceptual models based on incomplete understanding of concepts; "vernacular misconceptions," misunderstandings of words based on their common use versus their scientific use; and "factual misconceptions," incorrect or imprecise knowledge learned in early life that remains unchallenged (NRC 1997, p. 28). Misconceptions must be addressed and dismantled in order for students to reconstruct their knowledge, and therefore teachers should be prepared to take the following steps:

- *Identify students' misconceptions.*

- *Provide a forum for students to confront their misconceptions.*

- *Help students reconstruct and internalize their knowledge, based on scientific models. (NRC 1997, p. 29)*

Keeley and Harrington (2010) recommend using diagnostic tools such as probes and formative assessment to identify and confront student misconceptions and begin the process of reconstructing student knowledge. Keeley and Harrington's *Uncovering Student Ideas in Science* series contains probes targeted toward uncovering student misconceptions in a variety of areas. In particular, Volumes 1 and 2 of *Uncovering Student Ideas in Physical Science* (Keeley and Harrington 2010, 2014), about force/motion may be useful resources for addressing student misconceptions in this module.

Some commonly held misconceptions specific to lesson content are provided with each lesson so that you can be alert for student misunderstanding of the science concepts presented and used during this module. The American Association for the Advancement of Science has also identified misconceptions that students frequently hold regarding various science concepts (see the links at *http://assessment.aaas.org/topics*).

SRL PROCESS COMPONENTS

Table 3.2 illustrates some of the activities in the Amusement Park of the Future module and how they align to the SRL process before, during, and after learning.

Table 3.2. SRL Process Components

Learning Process Components	Example From Amusement Park of the Future Module	Lesson Number and Learning Component
BEFORE LEARNING		
Motivates students	Students watch a film with extreme amusement rides and evaluate the physical factors that make it exciting.	Lesson 1, Introductory Activity/ Engagement
Evokes prior learning	Students tap into their prior experience with amusement rides and recall the ways they moved in the ride.	Lesson 1, Introductory Activity/ Engagement
DURING LEARNING		
Focuses on important features	Students do a jigsaw activity with guidance from the teacher to form groups on • Spinning • Height • Feeling of weightlessness • Speed	Lesson 2, Activity/Exploration
Helps students monitor their progress	Students record findings for research on the speed and energy of various well-known amusement rides in their STEM Research Notebook. Teachers provide feedback on their completeness and accuracy.	Lesson 2, Elaboration/Application of Knowledge
AFTER LEARNING		
Evaluates learning	In the final challenge, groups present their final report on their new amusement park of the future project and receive feedback from the teacher and community members.	Lesson 3, Activity/Exploration
Takes account of what worked and what did not work	In the final challenge, students are asked to reflect on the feedback they receive on their group project and offer a plan of action for a re-design.	Lesson 3, Activity/Exploration

STRATEGIES FOR DIFFERENTIATING INSTRUCTION WITHIN THIS MODULE

For the purposes of this curriculum module, differentiated instruction is conceptualized as a way to tailor instruction—including process, content, and product—to various student needs in your class. A number of differentiation strategies are integrated into lessons across the module. The problem- and project-based learning approach used in the lessons is designed to address students' multiple intelligences by providing a variety of entry points and methods to investigate the key concepts in the module. Differentiation strategies for students needing support in prerequisite knowledge can be found in Table 3.1 (p. 26). You are encouraged to use information gained about student prior knowledge during introductory activities and discussions to inform your instructional differentiation. Strategies incorporated into this lesson include flexible grouping, varied environmental learning contexts, assessments, compacting, and tiered assignments and scaffolding.

Flexible Grouping. Students work collaboratively in a variety of activities throughout this module. Grouping strategies you might employ include student-led grouping, grouping students according to ability level, grouping students randomly, or grouping them so that students in each group have complementary strengths (for instance, one student might be strong in mathematics, another in art, and another in writing). You may also choose to group students based on their interest in different types of amusements when conducting historical research for the time line. For Lesson 2, you may choose to maintain the same student groupings as in Lesson 1 or regroup students according to another of the strategies described here. You may therefore wish to consider grouping students in Lesson 2 into design teams for the trebuchet. For Lesson 3, grouping should be based on the types of amusements students wish to design for the park. Place students who want to design different amusements together so that the park they develop as a group has more variety.

Varied Environmental Learning Contexts. Students have the opportunity to learn in various contexts throughout the module, including alone, in groups, in quiet reading and research-oriented activities, and in active learning through inquiry and design activities. In addition, students learn in a variety of ways, including through doing inquiry activities, journaling, reading fiction and nonfiction texts, watching videos, participating in class discussion, and conducting web-based research.

Assessments. Students are assessed in a variety of ways throughout the module, including individual and collaborative formative and summative assessments. Students have the opportunity to produce work via written text, oral and media presentations, and modeling. You may choose to provide students with additional choices of media for their products (for example, PowerPoint presentations, posters, or student-created websites or blogs).

Compacting. Based on student prior knowledge, you may wish to adjust instructional activities for students who exhibit prior mastery of a learning objective. For instance, if

some students exhibit mastery of energy transfer in Lesson 2, you may wish to limit the amount of time they spend practicing these skills and instead introduce ELA or social studies connections with associated activities.

Tiered Assignments and Scaffolding. Based on your awareness of student ability, understanding of concepts, and mastery of skills, you may wish to provide students with variations on activities by adding complexity to assignments or providing more or fewer learning supports for activities throughout the module. For instance, some students may need additional support in identifying key search words and phrases for web-based research or may benefit from cloze sentence handouts to enhance vocabulary understanding. Other students may benefit from expanded reading selections and additional reflective writing or from working with manipulatives and other visual representations of mathematical concepts. You may also work with your school librarian to compile a set of topical resources at a variety of reading levels.

STRATEGIES FOR ENGLISH LANGUAGE LEARNERS

Students who are developing proficiency in English language skills require additional supports to simultaneously learn academic content and the specialized language associated with specific content areas. WIDA has created a framework for providing support to these students and makes available rubrics and guidance on differentiating instructional materials for English language learners (ELLs) (see *www.wida.us/get.aspx?id=7*). In particular, ELL students may benefit from additional sensory supports such as images, physical modeling, and graphic representations of module content, as well as interactive support through collaborative work. This module incorporates a variety of sensory supports and offers ongoing opportunities for ELL students to work collaboratively. The focus in this module on amusement parks in a global context affords opportunities to access the culturally diverse experiences of ELL students in the classroom.

Teachers differentiating instruction for ELL students should carefully consider the needs of these students as they introduce and use academic language in various language domains (listening, speaking, reading, and writing) throughout this module. To adequately differentiate instruction for ELL students, teachers should have an understanding of the proficiency level of each student. WIDA provides an assessment tool to help teachers assess English language proficiency levels at *www.wida.us/assessment/ ACCESS20.aspx*. The following five overarching WIDA learning standards are relevant to this module:

- Standard 1: Social and Instructional language. Focus on social behavior in group work and class discussions.

- Standard 2: The language of Language Arts. Focus on forms of print, elements of text, picture books, comprehension strategies, main ideas and details, persuasive language, creation of informational text, and editing and revision.

- Standard 3: The language of Mathematics. Focus on numbers and operations, patterns, number sense, measurement, and strategies for problem solving.

- Standard 4: The language of Science. Focus on safety practices, magnetism, energy sources, scientific process, and scientific inquiry.

- Standard 5: The language of Social Studies. Focus on change from past to present, historical events, resources, transportation, map reading, and location of objects and places.

SAFETY CONSIDERATIONS FOR THE ACTIVITIES IN THIS MODULE

All laboratory occupants must wear safety glasses or goggles during all phases of inquiry activities (setup, hands-on investigation, and takedown). In this module, building the Rube Goldberg machine and the trebuchet will likely require a hot glue gun, and teachers should instruct students on proper use and storage to avoid burns or possible lighting of fires. For more general safety guidelines, see the Safety in STEM section in Chapter 2 (p. 18).

Internet safety is also important. The teacher should develop an internet/blog protocol with students if guidelines are not already in place. Since students will use the internet for their research to acquire the needed data, the teacher should monitor students' access to ensure that they are accessing only websites that are clearly identified by the teacher. Further, the teacher should inform parents or guardians that students will create online multimedia presentations of their research and that these projects will be closely monitored by the teacher. It is recommended that the teacher not allow any website posts created by students to go public without being approved first by the teacher.

DESIRED OUTCOMES AND MONITORING SUCCESS

This module is divided into three lessons. In Lesson 1, the goals include an understanding of the physical factors that cause thrills in amusement rides, the way innovations in amusement park design have increased the level of thrills, and the psychology behind why we find such experiences thrilling. The goals of Lesson 2 are an understanding of energy and how it is transferred in machines. Lesson 3 focuses on a total package of design elements for an amusement park, including rides, refreshments, lines, parking, and restrooms, and students work collaboratively in a group. The desired outcomes for this module are outlined in Table 3.3 (p. 32), along with suggested ways to gather evidence to monitor student success. For more specific details on desired outcomes, see the Established Goals and Objectives section for the module (p. 23) and for the individual lessons.

Table 3.3. Desired Outcomes and Evidence of Success in Achieving Identified Outcomes

Desired Outcome	Evidence of Success	
	Performance Tasks	Other Measures
Students work in teams to develop a comprehensive timeline, including history, innovations, and psychology behind amusements.	Students are assessed on the following using project rubrics that focus on content and application of skills related to academic content: • Accurate interpretation of research • Accurate labeling of time frames for each factor • Integration of factors along timeline • Explanation of related events in timeline	Students maintain STEM Research Notebooks to reflect on strategies that might work as they design their amusements of the future.
Students work in teams to develop a Rube Goldberg machine design and analysis.	Students are assessed on the following using project rubrics that focus on content and application of skills related to academic content: • Building of Rube Goldberg machine given design specifications • Analysis of energy transfer in machine • Analysis of potential and kinetic energy in machine	
Students work in teams to develop a trebuchet design and analysis, using proper safety procedures.	Students are assessed on the following using project rubrics that focus on content and application of skills related to academic content: • Building of trebuchet given design specifications • Conversion of scale measurements to actual measurements • Analysis of energy transfer • Analysis of potential and kinetic energy in machine	
Students work in teams to develop a business plan.	Students are assessed on the following using project rubrics that focus on content and application of skills related to academic content: • Working in groups collaboratively • Creating an accurate scaled blueprint • Creating an accurate cost-benefit analysis • Conducting a community impact study and communicating results • Creating a marketing plan that is reasonable, with accompanying infomercial	

ASSESSMENT PLAN OVERVIEW AND MAP

The assessment plan is created with a suite of formative and summative assessments designed to support student work in the final challenge. Students examine the factors of designing an amusement park through various disciplinary lenses, including physics, engineering, mathematical modeling of energy, language arts (through marketing), psychology of thrills, history of amusement, environmental sustainability, finance, and community impact. Table 3.4 provides an overview of the major group and individual *products* and *deliverables*, or things that constitute the assessment for this module, such as the time line, Rube Goldberg machine, trebuchet, and amusement park presentation. See Table 3.5 (p. 34) for a full assessment map of formative and summative assessments in this module.

Table 3.4. Major Products and Deliverables in Lead Disciplines for Groups and Individuals

Lesson	Major Group Products and Deliverables	Major Individual Products and Deliverables
1	• Jigsaw activity on physics of amusements	• Presentation of ideas about how amusements have developed over time, using a timeline as a communication tool • STEM Research Notebook prompts
2	• Group participation in investigations (students are responsible for their own analyses and communication of the results)	• Presentation of scientific explanations of the Rube Goldberg machine • Trebuchet energy analysis • STEM Research Notebook prompts
3	• Group presentation of the business plan for a collaborative amusement park	• Development of an individual amusement for the collaborative park • STEM Research Notebook prompts

Table 3.5. Assessment Map for Amusement Park of the Future Module

Lesson	Assessment	Group/ Individual	Formative/ Summative	Lesson Objective Assessed
1	STEM Research Notebook *prompts*	Group / individual	Formative	• Determine the types of physical characteristics (dropping, spinning, traveling at great heights) that amusement rides use to create thrills in people.
1	Argumentation *graphic organizer*	Group/ individual	Formative	• Connect psychology research regarding what people experience on amusement rides with the history of amusement rides.
1	Performance *rubric*	Group	Formative	• Determine the types of physical characteristics (dropping, spinning, traveling at great heights) that amusement rides use to create thrills in people.
1	Participation *rubric*	Group	Formative	• Determine the types of physical characteristics (dropping, spinning, traveling at great heights) that amusement rides use to create thrills in people.
1	Timeline *rubric*	Group	Formative	• Create a timeline of one type of amusement ride or game and document its history and how it has changed over time.
1	Narrative *rubric*	Group	Formative	• Create a timeline of one type of amusement ride or game and document the history of the ride/ game and its change over time.
2	STEM Research Notebook *prompts*	Group/ individual	Formative	• Explain the sustainability issues involved in running an amusement park. • Compile costs and incomes of the business of amusement parks.
2	Rube Goldberg Machine *rubric*	Group	Summative	• Explain transfer from one type of energy to another (kinetic and potential). • Measure and graph kinetic energy of a moving object. • Create Rube Goldberg machine from a minimum of 3 different simple machines.

Table 3.5. (*continued*)

Lesson	Assessment	Group/ Individual	Formative/ Summative	Lesson Objective Assessed
2	Engineering Design Process *scoring guide* Kinetic and Potential Energy Calculations from Trebuchet *scoring guide*	Group	Formative	• Build a working trebuchet to launch gumdrops. • Measure and graph kinetic energy of a moving object.
3	STEM Research Notebook *prompts*	Group/ individual	Formative	• Describe components of amusement park project.
3	Presentation and Report *rubric*	Group	Summative	• Draw a scaled blueprint of the amusement park with group members, taking into consideration foot traffic and refreshment issues. • Conduct a cost-benefit analysis for the amusement park. • Conduct an impact study for the proposed amusement park. • Create a marketing plan and infomercial for the proposed amusement park. • Write a report including blueprint, scale prototype drawing or mock-up, cost analysis, impact study, and marketing plan for the ride or game.

MODULE TIMELINE

The module can be described as three segments of work: (1) investigations involving the history and psychology of amusements; (2) exploration of the physical principles behind amusements; and (3) design and development of the amusement park prototype, with accompanying business plan. Tables 3.6–3.10 (pp. 37–38) provide lesson timelines for each week of the module.

Table 3.6. STEM Road Map Module Schedule for Week One

Day 1	Day 2	Day 3	Day 4	Day 5
Lesson 1 *The Thrill of the Ride— History and Psychology of Amusement Parks*	*Lesson 1* *The Thrill of the Ride— History and Psychology of Amusement Parks*	*Lesson 1* *The Thrill of the Ride— History and Psychology of Amusement Parks*	*Lesson 1* *The Thrill of the Ride— History and Psychology of Amusement Parks*	*Lesson 1* *The Thrill of the Ride— History and Psychology of Amusement Parks*
• Students launch the module by having students study close-up photos of people riding amusements to discuss why people seek thrill rides and establish extreme versions of different rides, examining the types of motion the rides offer.	• Students investigate amusement parks throughout history, beginning with London World's Fair in 1851 and Coney Island in 1880, and make predictions for amusements in the future.	• Students each choose to focus on one type of amusement (roller coasters, height amusements, spinning amusements, or games) and work as a class to create an elaborated timeline. • Students present findings in a gallery walk.	• Students research psychological reasons people seek thrills and amusements. • Students synthesize their ideas to show major themes and connect the themes to evidence from their research.	• Students synthesize their elaborated timelines with their findings from the research on the psychology of amusement rides and present to the whole class and to outside reviewer.

Table 3.7. STEM Road Map Module Schedule for Week Two

Day 6	Day 7	Day 8	Day 9	Day 10
Lesson 2 *Faster, Higher, and Safer*	*Lesson 2* *Faster, Higher, and Safer*	*Lesson 2* *Faster, Higher, and Safer*	*Lesson 2* *Faster, Higher, and Safer*	*Lesson 2* *Faster, Higher, and Safer*
• Students investigate various types of energy transfer and share with the whole class their research on roller coasters that claim to be the highest, fastest, and steepest. • Students also examine roller coaster simulators to play with variables and observe their outcomes.	• Students build a Rube Goldberg machine and calculate kinetic energy in different energy transfer scenarios, graph results, and share with the whole class.	• Students begin plans for building a trebuchet.	• Students start to build the trebuchet and calculate the kinetic and potential energy in the systems.	• Students finish building the trebuchet and calculate the kinetic and potential energy in the systems. • Students create graphs to represent their results and develop arguments for their design decisions.

Table 3.8. STEM Road Map Module Schedule for Week Three

Day 11	Day 12	Day 13	Day 14	Day 15
Lesson 2 *Faster, Higher, and Safer* • Students begin to research and compile safety factors for their designed amusement park ride or game.	*Lesson 2* *Faster, Higher, and Safer* • Students continue to research and compile safety factors.	*Lesson 3* *Amusement Park of the Future Design Challenge* • Students work on scale drawings of a chosen ride from a local amusement park, noting design features of types of rides, games, refreshments, and areas for lining up.	*Lesson 3* *Amusement Park of the Future Design Challenge* • Groups work on blueprint for amusement park.	*Lesson 3* *Amusement Park of the Future Design Challenge* • Groups complete blueprint for amusement park.

Table 3.9. STEM Road Map Module Schedule for Week Four

Day 16	Day 17	Day 18	Day 19	Day 20
Lesson 3 *Amusement Park of the Future Design Challenge* • Groups start to make scale prototype for amusement park.	*Lesson 3* *Amusement Park of the Future Design Challenge* • Groups continue to work on scale prototype for amusement park.	*Lesson 3* *Amusement Park of the Future Design Challenge* • Groups complete scale prototype and present it for a peer review.	*Lesson 3* *Amusement Park of the Future Design Challenge* • Groups work on cost-benefit analysis for amusement park.	*Lesson 3* *Amusement Park of the Future Design Challenge* • Groups complete marketing plan for amusement park.

Table 3.10. STEM Road Map Module Schedule for Week Five

Day 21	Day 22	Day 23	Day 24	Day 25
Lesson 3 *Amusement Park of the Future Design Challenge* • Groups develop infomercial for amusement park.	*Lesson 3* *Amusement Park of the Future Design Challenge* • Groups work on integrating and polishing pieces of their challenge product: blueprints, prototype, cost-benefit analysis, impact study, marketing plan, and infomercial.	*Lesson 3* *Amusement Park of the Future Design Challenge* • Groups begin to present blueprints, prototype, cost-benefit analysis, impact study, marketing plan, and infomercial. • Students, teacher, and community members document strengths and weaknesses for future discussion.	*Lesson 3* *Amusement Park of the Future Design Challenge* • Groups continue to present blueprints, prototype, cost-benefit analysis, impact study, marketing plan, and infomercial. • Students, teacher, and community members document strengths and weaknesses for future discussion.	*Lesson 3* *Amusement Park of the Future Design Challenge* • Whole class engages in discussion and analysis of strengths and weaknesses of each group's challenge product. • Groups meet to improve and adapt plan based on discussion.

RESOURCES

Teachers have the option to coteach portions of this module and may want to combine classes for activities such as mathematical modeling, geometric investigations, discussing social influences, or conducting research. The media specialist can help teachers locate resources for students to view and read about the history of amusements and provide technical help with spreadsheets, timeline software, and multimedia production software. Special educators and reading specialists can help find supplemental sources for students needing extra support in reading and writing. Additional resources may be found online. Community resources for this module may include town council or business bureau members for hearing the business plan presentations, school administrators, and parents.

REFERENCES

Johnson, C. C., T. J. Moore, J. Utley, J. Breiner, S. R. Burton, E. E. Peters-Burton, J. Walton, and C. L. Parton. 2015. The STEM Road Map for grades 6–8. In *STEM Road Map: A framework for integrated STEM education,* ed. C. C. Johnson, E. E. Peters-Burton, and T. J. Moore, 96–123. New York: Routledge. *www.routledge.com/products/9781138804234.*

Keeley, P., and R. Harrington. 2010. *Uncovering student ideas in physical science. Vol. 1, 45 new force and motion assessment probes.* Arlington, VA: NSTA Press.

Keeley, P., and R. Harrington. 2014. *Uncovering student ideas in physical science. Vol. 2, 39 new electricity and magnetism formative assessment probes.* Arlington, VA: NSTA Press.

National Research Council (NRC). 1997. *Science teaching reconsidered: A handbook.* Washington, DC: National Academies Press.

Peters-Burton, E. E., P. Seshaiyer, S. Burton, J. Drake-Patrick, and C. C. Johnson. 2015. The STEM road map for grades 9–12. In *The STEM road map: A framework for integrated STEM Education,* ed. C. C. Johnson, E. E. Peters-Burton, and T. J. Moore, 124–162. New York: Routledge Publishing. *www.routledge.com/products/9781138804234.*

AMUSEMENT PARK OF THE FUTURE LESSON PLANS

Erin Peters-Burton, Janet B. Walton, and Carla C. Johnson

Lesson Plan 1: The Thrill of the Ride—History and Psychology of Amusement Parks

This lesson launches the module by having the students examine photos, watch a video of an art project with extreme amusement rides, and view the accompanying website, which connects how particular characteristics of physics and engineering are used in amusements to generate emotions in people such as fear, excitement, and exhilaration. Students then elaborate on this baseline knowledge to investigate the history of amusements, note improvement to technologies and innovation, and connect the types of amusements with human psychological triggers.

ESSENTIAL QUESTIONS

- Why do people seek thrills in amusement parks?

- What kinds of characteristics do engineers, physicists, and mathematicians use in the design of amusement rides to make the largest impact on people?

ESTABLISHED GOALS AND OBJECTIVES

At the conclusion of this lesson, students will be able to do the following:

- Determine the types of physical characteristics (dropping, spinning, traveling at great heights) that amusement rides use to create thrills in people.

- Create a timeline of one type of amusement ride or game and document its history and how it has changed over time.

- Connect psychology research regarding what people experience on amusement rides with the history of amusement rides.

TIME REQUIRED

- 5 days (approximately 45 minutes each day; see Table 3.6, p. 37)

MATERIALS

- STEM Research Notebooks (1 per student; see p. 25 for STEM Research Notebook student handout)

- Computers with internet access

- Timeline creation software or paper, markers, and meter sticks

- Safety glasses or goggles

- Ball on a string

- Sturdy paper plate with a rim so that a marble can spin around the plate edge

- Scissors

- Marble

CONTENT STANDARDS AND KEY VOCABULARY

Table 4.1 lists the content standards from the *Next Generation Science Standards* (*NGSS*), *Common Core State Standards* (*CCSS*), and the Framework for 21st Century Learning that this lesson addresses, and Table 4.2 (p. 44) presents the key vocabulary. Vocabulary terms are provided for both teacher and student use. Teachers may choose to introduce some or all of the terms to students.

Table 4.1. Content Standards Addressed in STEM Road Map Module Lesson 1

NEXT GENERATION SCIENCE STANDARDS

PERFORMANCE EXPECTATIONS

The Performance Expectations for this lesson are not directly assessed; rather, engagement and foundational physics concepts are developed in this lesson. The Performance Expectations for the module are directly assessed in Lessons 2 and 3.

COMMON CORE STATE STANDARDS FOR MATHEMATICS

MATHEMATICAL PRACTICES

- MP1. Make sense of problems and persevere in solving them.
- MP2. Reason abstractly and quantitatively.
- MP3. Construct viable arguments and critique the reasoning of others.
- MP5. Use appropriate tools strategically.

MATHEMATICAL CONTENT

- 6.SP.B.5b. Describing the nature of the attribute under investigation, including how it was measured and its units of measurement.

COMMON CORE STATE STANDARDS FOR ENGLISH LANGUAGE ARTS

READING STANDARDS

- RI.6.1. Cite textual evidence to support analysis of what the text says explicitly as well as inferences drawn from the text.
- RI.6.4. Determine the meaning of words and phrases as they are used in a text, including figurative, connotative, and technical meanings.
- RI.6.7. Integrate information presented in different media or formats (e.g., visually, quantitatively) as well as in words to develop a coherent understanding of a topic or issue.

WRITING STANDARDS

- W.6.1. Write arguments to support claims with clear reasons and relevant evidence.
- W.6.2. Write informative/explanatory texts to examine a topic and convey ideas, concepts, and information through the selection, organization, and analysis of relevant content.

SPEAKING AND LISTENING STANDARDS

- SL.6.1. Engage effectively in a range of collaborative discussions (one-on-one, in groups, and teacher-led) with diverse partners on grade 6 topics, texts, and issues, building on others' ideas and expressing their own clearly.
- SL.6.2. Interpret information presented in diverse media and formats (e.g., visually, quantitatively, orally) and explain how it contributes to a topic, text, or issue under study.

LITERACY STANDARDS

- L.6.1. Demonstrate command of the conventions of standard English grammar and usage when writing or speaking.

Table 4.1. (*continued*)

FRAMEWORK FOR 21ST CENTURY LEARNING
Creativity and Innovation, Critical Thinking and Problem Solving, Communication and Collaboration, Information Literacy, Media Literacy, ICT Literacy, Flexibility and Adaptability, Initiative and Self-Direction, Social and Cross Cultural Skills, Productivity and Accountability, Leadership and Responsibility

Table 4.2. Key Vocabulary in Lesson 1

Key Vocabulary	Definition
acceleration	a change in an object's velocity, which could be a change in how fast it is moving or a change in direction
centrifuge	a machine that causes material (or a human) to travel in a circle
centripetal force	the force that keeps an object rotating in a circle
energy	the capacity for doing work; can be different types such as chemical, electric, mechanical; can also be kinetic or potential
force	a push or a pull on an object
gravity, or g-force	the force on an object as a result of acceleration or gravity measured in magnitude of Earth's gravity; the g-force increases when acceleration increases, which may be due to an increase in speed reached per unit of time or a decrease in the amount of time needed for the speed to be reached (g-force = acceleration from motion/acceleration from gravity)
mass	the amount of material in a body; it determines the amount of weight a body has due to the pull of gravity
psychology	the study of the human mind and behavior
speed	the rate at which something moves or covers distance; the distance a moving object travels per unit of time (if object 1 travels farther than object 2 over the same time, then object 1 has greater speed)
velocity	description of an object's motion that includes how fast it goes and the direction it is going
work	when force being applied to an object moves the object over a distance

4

TEACHER BACKGROUND INFORMATION

The extreme g-forces proposed in the amusement rides in this lesson, such as 17 g, cannot be tolerated by humans. Astronauts who experience 3 g and changes in forces during launch and reentry must train for this experience for months in advance. This website provides real-life data on amusement park rides and g-forces, along with some explanation of physics experiments that can be done at the popular Physics Days at amusement parks: *http://physicsbuzz.physicscentral.com/2013/04/roller-coaster-g-forces-weve-got-data.html*. Teachers who have extra resources may want to consider organizing a Physics Day at a local amusement park. Table 4.3 gives some examples of g-forces.

Table 4.3. Typical Examples of G-Forces

Example	G-Force
Standing on Earth at sea level (standard)	1 g
Saturn V moon rocket just after launch	1.14 g
Going from 0 to 100 kilometers per hour in 2.4 seconds	1.55 g
Space Shuttle, maximum during launch and reentry	3 g

COMMON MISCONCEPTIONS

Students will have various types of prior knowledge about the concepts introduced in this lesson. Table 4.4 outlines some common misconceptions students may have concerning these concepts. Because of the breadth of students' experiences, it is not possible to anticipate every misconception that students may bring as they approach this lesson. Incorrect or inaccurate prior understanding of concepts can influence student learning in the future, however, so it is important to be alert to misconceptions such as those presented in the table.

Table 4.4. Common Misconceptions About the Concepts in Lesson 1

Topic	Student Misconception	Explanation
Motion: Speed, velocity, and acceleration are three concepts that describe the motion of an object.	Speed and velocity are the same.	Speed is a quantity (number) that only describes the distance traveled over time. Velocity is also a quantity, like speed, but also describes direction.
	Acceleration is defined only as an increase in speed or as a change in speed.	Acceleration is a change in velocity, so acceleration could be a change in the speed of an object or the change in direction of an object traveling at the same speed, or both.

PREPARATION FOR LESSON 1

Review the Teacher Background Information provided, and preview the websites suggested in the Learning Plan Components section below to ensure that you have foundational knowledge in the ways that professional amusement designers use physics and psychology to create amusements. In this lesson, you will observe student interest to form groups for the design challenge. Teacher cues are given within the lesson plan to help you consider variables when forming groups. Ultimately, the challenge project groups should include students who have a variety of amusement interests within the same theme. For example, if students want to create a park with a haunted theme, the group should consist of students who want spooky-themed rides, with one or two students focusing on roller coasters, one or two on a height-related ride, one or two on spinning rides, and one or two on dart- or ball-throwing games. The final product will be a complete park with a full complement of amusements.

Have your students set up their STEM Research Notebooks (see pp. 24–25 for discussion and student instruction handout).

LEARNING PLAN COMPONENTS
Introductory Activity/Engagement

Connection to the Challenge: Begin each day of this lesson by directing students' attention to the driving question for the module and challenge: How can we use what we know about the development of amusements, the ways people experience thrills, and the laws of physics to propose new amusements that are both safe and extreme? Ask them why they think amusement parks are so popular. What makes them thrilling? Hold a brief student discussion of how their learning in the previous days' lessons contributed to their ability to create their plan for their innovation in the final challenge. You may wish to create a class list of key ideas on chart paper or the board or have students create a STEM Research Notebook entry with this information.

The introductory activity is multipronged. In science class, students watch an art video of a fictional research project that designs and creates extreme amusement rides. In mathematics, students examine the accompanying website. Although the actual research project is fictional, the extreme nature of the rides will engage students, and the discussion in the video pinpoints examples of physical principles that are key to designing successful amusement rides. In social studies class, students examine photos and use historical inquiry to make inferences about the emotions and motivation of people while riding amusements.

Science Class: Watch the 6½-minute video on the Centrifuge Brain Project (visit YouTube and search for "Centrifuge Brain Project" or access the video directly at *www.youtube.com/*

watch?v=RVeHxUVkW4w). As described earlier, this video is an art project that emphasizes important features of amusements and why people design them.

STEM Research Notebook Prompt

Ask students the following questions after they watch the video. Have them initially record their answers in their STEM Research Notebooks, then discuss as a class.

- What things do you think were the most important in this video?

- What characteristics did the researchers investigate? (Examples include acceleration, spinning, height, feeling of weightlessness, and speed.)

- What kinds of experiences did the designers try to simulate? What was their inspiration?

- The video mentions "6 g's." What do you think that means?

- What does the scientist mean by "gravity is a mistake"?

- Why do you think people go to amusement parks and ride the rides?

- Would you ride these rides? Why or why not?

- Do you think they are real? Why or why not?

Explain to the class that this video and accompanying website are actually an art project, but they are based on science, engineering, and mathematics.

Mathematics Connection: Have students investigate the website of the Institute for Centrifugal Research, creator of the Centrifuge Brain Project: *www.icr-science.org/index.htm*.

STEM Research Notebook Prompt

Students should respond to the following prompt in their Research Notebooks: Choose one of the rides and make estimates to determine the speed (distance divided by time) that the people are traveling. Record all ideas and research in your notebook.

Social Studies Class: Have students look at the pictures of roller coaster riders' faces on this website: *www.thisiscolossal.com/2010/10/magic-feelings-portraits-of-roller-coaster-riders*.

STEM Research Notebook Prompt

Ask students to write their responses to the following questions in their Research Notebook, and then discuss as a class:

- What do you think these people are feeling? Why?

- What do you think they are doing? Explain.

- Why do people seek thrills?

- Why are amusement parks popular ways to spend leisure time?

- In what ways do people connect to each other as a community at amusement parks?

- Is there a culture of amusement parks?

- How can science and engineering improve a person's experience in an amusement park?

Have a class discussion about the experiences that students have had in amusement parks, including both rides and games that they have played. Ask them to compare their experiences with what they thought the people in the photos were feeling and why people go to amusement parks.

English Language Arts (ELA) Connection: Ask students to interpret the technical information given in the video and on the websites that they explored in science, mathematics, and social studies class. Provide support in any areas of need, such as explaining technical language or clearing up any misinterpretations of the information communicated.

Activity/Exploration

Science Class: Use a jigsaw activity to have students investigate and elaborate on the physics involved in the project. In a jigsaw activity, students are grouped in two different ways. Here, they will be divided first into expert groups and then into home groups. Divide the class evenly into four expert groups that will research different types of motion of amusement rides: one group for spinning, one for height, one for feeling of weightlessness, and one for speed. Ask students to use the internet to investigate what concepts are involved in their group's assigned type of motion.

Once you have assessed that groups have compiled enough information, form home groups that each consist of one student who is a spinning expert, one height expert, one feeling of weightlessness expert, and one speed expert. Have these experts teach their groups about the key features of their concepts. You may wish to create an individual assessment of all four concepts to ensure interdependence in the group.

STEM Research Notebook Prompt

Students should record all information they researched for their expert portion of the jigsaw activity, as well as what they learn from their home group members, in their STEM Research Notebooks. Then, have students use the internet to investigate the history of amusements, beginning with the London World's Fair in 1851, Coney Island in the 1880s, and the New York World's Fair in 1939–1940, with a focus on the physics behind the rides and games. Students should use the concepts they explored in the jigsaw activity to identify the physics concepts associated with the rides and games. Games explored here should include some physical motion. Ask students to record their findings from this research in their STEM Research Notebooks. The following websites represent a core beginning for online research on the topic.

- Amusement park features

    - *www.themeparktourist.com/features*

- History of roller coasters

    - *www.pbs.org/wgbh/amex/coney/sfeature/history.html*

    - *www.aceonline.org/coasterhistory*

    - *www.ultimaterollercoaster.com/coasters/history/start*

- Coney Island history

    - *www.coneyislandhistory.org*

    - *www.history.com/topics/new-york-city/videos/coney-island-ups-and-downs*

- Early world's fairs

    - *www.1939nyworldsfair.com/worlds_fair/wf_tour/zone-7/zone-7.htm*

    - *www.boweryboyshistory.com/2015/03/the-crystal-palace-americas-first-worlds-fair-and-bizarre-treasures-of-the-19th-century.html*

Ask students to choose one type of amusement (roller coasters, height amusements, spinning amusements, or dart- or ball-throwing games) and develop a timeline from the 1880s to present day, noting the changes in physics concepts, such height, speed, or direction, over time. Students may use timeline creation software or create their time-lines on paper using markers and meter sticks. Safety note: Use of meter sticks requires eye protection.

Students should first take notes from the websites regarding their amusement of choice. For example, if students choose a height-related amusement, they would look at each website for this amusement, noting the date of the amusement (or approximating the date) and the way people are placed in the ride, safety features, how the mechanics of the ride make it go up and down and how high the people go in the ride. Students should then compose their notes into a table with dates and important physics features, which they can then put into the timeline software or create a paper-and-pencil ver-sion of the timeline. After individual timelines are completed, students who choose the

same type of amusement should group together to compare their findings and create one group timeline with all compiled information. The whole class should then discuss their amusement timelines and the teacher facilitates the creation of a class timeline with all of the different types of amusements. Students should discuss as a class the trends that they see when all of the amusements are compiled and record the ideas in their notebook. For example, students may find that safety features have adapted over time, such as the progression from straps that passengers grip to seatbelts to hydraulic shoulder harnesses. Students may also notice that amusements have become taller and faster as technology has improved.

Mathematics Connection: Have students begin by researching the units of distance, time, speed, and acceleration in both the metric and standard systems. Lead students to see that the units for distance, time, speed, and acceleration are related. Distance divided by time is speed, and speed divided by time is acceleration. In addition to a change in speed, a change in the direction of the object causes acceleration. Have students research the concept of g-force (g), a measurement using a multiple of the acceleration due to gravity. Then, using the information they find, have them calculate 6 g.

Ask students the following questions:

- Is it possible for humans to experience an acceleration of 6 g?

- What g-force do astronauts experience when they take off out of the Earth's atmosphere?

Have students compare 6 g with the data on the graphs of a typical amusement park on this website: *http://physicsbuzz.physicscentral.com/2013/04/roller-coaster-g-forces-weve-got-data.html.*

ELA Connection: Continue to monitor students' correct interpretation of the technical information they identify in their research.

Social Studies Class: Have students read the article titled "Thrills & Chills" on the Psychology Today website, found at *www.psychologytoday.com/articles/199905/thrills-chills.* Ask them to write the major ideas presented in the article in their Research Notebook . Next, have students pair up with partners to compare the major ideas they found with the major ideas their partners found. Have students talk about the differences until they come to a consensus. Then, as a whole group, discuss the major ways that amusement park designers create an experience of thrills and chills for patrons.

STEM Research Notebook Prompt

Ask students to write their answers to the following in their Research Notebooks and discuss as a class:

- What did you find intriguing?
- What type of amusement park might you want to design in the future?
- What would be the themes of that park?

You can use students' responses to these questions to begin documenting student ideas to help inform groups for the final challenge. For example, some might want to design a park with a haunted house theme, while others might want to design an adventure park.

Explanation

Science Class: Facilitate a discussion of core physics concepts in the field of Newtonian mechanics. Key concepts are distance, time, speed, velocity, acceleration, mass, force, energy, and work. Besides what students have already learned so far about these concepts, they should know that mass and acceleration are related to force and that to do work, a force placed on an object must make it travel a distance.

- Next, facilitate a discussion showing that force can be applied in a straight line, demonstrating Newton's three laws of motion, but that force can also cause an object to travel in a circle, which is called centripetal force. Students may have heard of centrifugal force, but this is a fictitious force one uses to describe oneself as the object in motion when one is going in a circle. Actually, what is happening from an external frame of reference is that an object traveling in a circle will continue to try to go in a straight line (because of Newton's first law of motion, the law of inertia), but when centripetal force pulls in the object toward the center point of the circle, the object is forced into the path of a circle.

- You can demonstrate this by spinning a ball on a string above your head like a helicopter propeller. Then, carefully let go of the string and note that the ball no longer travels in a circle but continues in a straight line. Your students can experiment with this phenomenon by cutting a pie wedge out of a sturdy paper plate (about one-quarter of the plate will work), then spinning a marble around the rim on the inside of the plate. When the marble reaches the cutout part, it will immediately travel in a straight line. This demonstrates that the force is not within the marble (a common misconception) but is caused by the paper plate pushing the marble into a circle. *Safety note:* This activity requires eye protection (safety glasses or goggles).

Continue facilitating a discussion of the physics principles applied to the motion of amusement park rides and games. Ask students what distance, time, speed, acceleration, mass, force, energy, and work have to do with the different types of motion of amusement rides: spinning, height, feeling of weightlessness, and speed. Ask students

how they would maximize speed, acceleration, force, and height for the rides they have researched. Refer students to the timeline activity to help them think about how physical properties of amusement have changed over time and how they might adapt for the future.

Mathematics Connection: Have students use the physics concepts from their timelines to graph changes in measures over time, such as height, speed, or direction, to explain innovations in amusement rides and the rate at which these innovations occurred.

ELA Connection: Ask students to write a narrative from the perspective of one person during his or her life span (60–90 years) over a time period included on their timelines. The essay should focus on the innovations of amusement parks the person visited.

Social Studies Class: Students continue working on their timelines, adding the major U.S. historical events that occurred during the years covered by the timeline, including (but not limited to) the Great Depression and World Wars I and II.

Elaboration/Application of Knowledge

Science Class: Have students use the internet to research the psychology of amusement rides, especially thrill rides like roller coasters. In addition to the "Thrills & Chills" article students read earlier, the following websites will help them begin their search.

- Psychology of amusements
  - *www.bostonglobe.com/2014/06/20/why-you-love-amusement-park-games-and-rides/ MmuFsFi07pEH5AVM7YUasO/story.html*
  - *www.nytimes.com/1988/08/02/science/why-do-people-crave-the-experience.html*
  - *www.technologytell.com/gadgets/149668/geotech-4-psychological-tricks-behind-disneys-theme-park-success*

- Psychological effects of roller coasters
  - *www.sciencenewsforstudents.org/article/roller-coaster-thrills*
  - *www.sciencealert.com/the-science-behind-why-we-love-terrifying-ourselves-on-rollercoasters*
  - *www.huffingtonpost.ca/2015/08/27/why-people-love-roller-co_n_8050428.html*

Students can work in small groups to make claims about the psychological effects of amusement rides. The claims should be backed up by evidence. Distribute copies of the Argumentation Graphic Organizer found at the end of this lesson (see p. 56) to help students compile their claims, evidence, and reasoning.

Mathematics Connection: Students examine each other's timelines and graphs to note the most rapid rates of innovation of amusements and how they correspond to major U.S. events investigated in social studies. Ask students to pose ideas about why the periods of innovations happened when they did. For example, when soldiers returned from war, technological innovations from the war were incorporated into amusement parks. Additionally, people had more time and money to seek leisure.

ELA Connection: Check the claims, evidence, and reasoning on students' Argumentation Graphic Organizers to ensure that they contain appropriate information. Students should then present their claims with evidence and reasoning for peer review.

Social Studies Class: Students should incorporate their peer-reviewed claims about the psychology of amusements based on the dates of the studies they find in the internet search into their timelines they produced earlier on the history of amusement and changes in physical characteristics of amusements. For example, students may find that all of the claims about the psychology of amusements will occur at the end of the timeline, because psychology is a new field of study compared to physics. Students will use these elaborated timelines to reflect on and help plan their own amusement parks for the challenge, noting the possible changes over time and extrapolating forward from this information.

Evaluation/Assessment

This section of the lesson plan lists assessments that are administered throughout the lesson. They include both formative assessments that are used by the teacher to gather information about student learning and by the student as a mechanism for feedback about their learning. Students may be assessed on the following measures:

- Participation in whole-class discussions, making claims that are backed up by evidence and reasoning
- Creation of a timeline representing several lines of information simultaneously:
  - History of amusements
  - Physical improvements of amusements
  - Psychology findings about thrills and chills
  - Major historical events such as wars or innovations outside of the amusement business
- Creation of a narrative of one person's experience with amusements over a time period included on the timeline
- Performance Rubric

- Participation Rubric
- Timeline Rubric
- Narrative Rubric

INTERNET RESOURCES

Amusement park features
- *www.themeparktourist.com/features*

Coney Island history
- *www.coneyislandhistory.org*
- *www.history.com/topics/new-york-city/videos/coney-island-ups-and-downs*

Data on amusement park rides and g-forces
- *http://physicsbuzz.physicscentral.com/2013/04/roller-coaster-g-forces-weve-got-data.html*

Institute for Centrifugal Research, creator of the Centrifuge Brain Project
- *www.icr-science.org/index.htm*

Early world's fairs
- *www.1939nyworldsfair.com/worlds_fair/wf_tour/zone-7/zone-7.htm*
- *www.boweryboyshistory.com/2015/03/the-crystal-palace-americas-first-worlds-fair-and-bizarre-treasures-of-the-19th-century.html*

History of amusement park features
- *http://inventors.about.com/od/tstartinventions/ss/theme_park.htm*

History of roller coasters
- *www.pbs.org/wgbh/amex/coney/sfeature/history.html*
- *www.aceonline.org/coasterhistory*
- *www.ultimaterollercoaster.com/coasters/history/start*

Images of roller coaster riders' faces
- *www.thisiscolossal.com/2010/10/magic-feelings-portraits-of-roller-coaster-riders*

Psychological effects of roller coasters
- *www.sciencenewsforstudents.org/article/roller-coaster-thrills*

- *www.sciencealert.com/the-science-behind-why-we-love-terrifying-ourselves-on-rollercoasters*

- *www.huffingtonpost.ca/2015/08/27/why-people-love-roller-co_n_8050428.html*

Psychology of amusements
- *www.bostonglobe.com/2014/06/20/why-you-love-amusement-park-games-and-rides/MmuFsFi07pEH5AVM7YUasO/story.html*

- *www.nytimes.com/1988/08/02/science/why-do-people-crave-the-experience.html*

- *www.technologytell.com/gadgets/149668/geotech-4-psychological-tricks-behind-disneys-theme-park-success*

"Thrills and Chills" article on the *Psychology Today* website
- *www.psychologytoday.com/articles/199905/thrills-chills*

4

Name: _____

ARGUMENTATION GRAPHIC ORGANIZER

Problem/Question:

Original Claim:

REASONING

EVIDENCE

1	
2	
3	

RATIONALE

VALID

REBUTTAL

1	
2	
3	

Conclusion:

PERFORMANCE RUBRIC

Component	Does Not Yet Meet Expectations	Meets Expectations	Exceeds Expectations
Timeline product supported by jigsaw activity: physics concepts	The concepts of spinning, height, weight, feeling of weightlessness, speed, and acceleration are not addressed, are only partially addressed, or do not conform to the understanding of the phenomena.	Most of the concepts of spinning, height, weight, feeling of weightlessness, speed, and acceleration are addressed and conform to the understanding of the phenomena.	All the concepts of spinning, height, weight, feeling of weightlessness, speed, and acceleration are addressed and conform to the understanding of the phenomena.
Timeline product: communication	Timeline has only superficial explanation of amusement, or changes in physical characteristics in amusement are not represented.	Timeline has accurate explanation of amusement, and changes in physical characteristics in amusement are represented.	Timeline has accurate explanation of amusement, and changes in physical characteristics in amusement are represented in both written and mathematical forms.
Proper units of measure	Units of measure are not included on products.	Units of measure are included for all physical characteristics of products.	Units of measure are included for all physical characteristics of products.
Participation in class discussion	Does not participate in class discussion about thrills and chills.	Participates minimally in class discussion about the main ideas in the thrills and chills article.	Connects to other students' ideas when participating in class discussion of thrills and chills.
Claims about psychological effects of amusements	Makes claims about psychological effects without supporting them with evidence or references.	Makes claims about psychological effects and supports them with evidence from references.	Makes claims about psychological effects, supports them with evidence from references, and connects the claims and evidence with reasoning based on principles from science.
Narrative of one person's experience with amusements over a time period included on the timeline	Narrative does not fit with the events of the timeline or is poorly written.	Narrative fits with the events of the timeline and is well written.	Narrative fits with the events of the timeline, is well written, and includes physics and psychological principles.

COMMENTS:

PARTICIPATION RUBRIC

Component	Emerging	Competent	Advanced
Follows guidelines of intellectual discussion and is civil	Criticizes other people personally instead of being critical of ideas; does not use appropriate language	Challenges the idea without solid reasoning; uses appropriate language	Challenges the idea with solid reasoning; uses appropriate language; diverts any unproductive discussion
Makes claim	Claim unoriginal and indirectly related to topic	Claim original and indirectly related to topic	Claim original and directly related to topic
Uses reliable sources for evidence	Uses unreliable resources, such as a blog without references	Uses only the textbook as resource	Uses outside reliable resources, such as a scientific journal or .gov or .edu website
Presents appropriate level of evidence	Uses opinion-based evidence	Presents only one piece of researched evidence	Presents more than one piece of researched evidence
Responds to the content of the discussion	Does not respond or response is unrelated to claim	Response is indirectly associated with claim	Response is aligned with claim
Connects with what previous person said	Comments are unrelated to current discussion	Stays on topic but makes no connection with previous person's comments	Acknowledges previous person's idea and elaborates on what was said
Able to defend his or her claim or rebuttal	Has no response	Has a response but cannot back it up with evidence	Has a response and is able to back it up with further evidence
Uses appropriate reasoning	Reasoning disconnected from claim	Reasoning superficially connected to claim	Reasoning directly connects claim to evidence

COMMENTS:

TIMELINE RUBRIC

Component	Emerging	Competent	Advanced
Historical events	Some of the historical events are incorrectly described or located on timeline.	One of the historical events is incorrectly described or located on timeline.	All historical events are correctly described or located on timeline.
Physical improvements of amusements	Some of the facts about physical improvements are incorrectly described or located on timeline.	One of the facts about physical improvements is incorrectly described or located on timeline.	All facts about physical improvements correctly described or located on timeline.
Psychological aspects of amusements	Some of the facts about psychological aspects are incorrectly described or located on timeline.	One of the facts about psychological aspects is incorrectly described or located on timeline.	All facts about psychological aspects correctly described or located on timeline.
Major events in society	Some of the facts or events in society are incorrectly described or located on timeline.	One of the facts or events in society is incorrectly described or located on timeline.	All facts and events in society are correctly described or located on timeline.

COMMENTS:

NARRATIVE RUBRIC

Component	Emerging	Competent	Advanced
Has background information that sets the stage	Background information is missing, incomplete, or inaccurate.	Background information is clearly communicated and accurate.	Background information is clearly communicated and accurate, and it enhances the narrative.
Focuses on changes in amusements over time	Focus on changes in amusements over time is missing, incomplete, or inaccurate.	Focus on changes in amusements over time is clearly communicated and accurate.	Focus on changes in amusements over time is clearly communicated and accurate, and it enhances the narrative.
Has a conclusion that synthesizes the information presented	Conclusion is missing, incomplete, or inaccurate.	Conclusion is clearly communicated and synthesizes the information presented.	Conclusion is clearly communicated, synthesizes the information presented, and enhances the narrative.

COMMENTS:

Lesson Plan 2: Faster, Higher, and Safer

In this lesson, students investigate types of energy transfer and run simulations to consider design factors for amusements such as speed, height, and sustainability. To get a sense of the physics behind the amusements, students will participate in two design processes. Students will design a Rube Goldberg machine to learn more about how simple machines propel an object across distances. Students will also design a trebuchet to learn more about projecting objects through space. They will use these common experiences as a launching pad for the design of their amusement. Students also gather information to make decisions on cost and safety to balance the practical aspects of amusements (safety, cost, sustainability) with the fun features of amusements (thrills due to speed, height, and spinning). Finally, students apply this knowledge to the design of an amusement ride or dart- or ball-throwing game and make scale drawings of their amusements.

ESSENTIAL QUESTIONS

- What types of energy are involved with amusement rides that go extremely fast and whip around sharp angles?

- How do you get amusement rides such as roller coasters to move quickly to provide thrills and still be safe for the riders?

- What ecological impact do amusement parks have on the community?

ESTABLISHED GOALS AND OBJECTIVES

At the conclusion of this lesson, students will be able to do the following:

- Distinguish the different types of energy (e.g., mechanical, , sound, heat) and explain transfer from one type of energy to another (kinetic and potential).

- Measure and graph kinetic energy of a moving object.

- Explain the sustainability issues involved with running an amusement park.

- Use a computer simulation to successfully balance the thrill of an amusement ride (speed, spin) with the safety of the rider so that the ride is compelling but safe.

- Compile costs and incomes of the business of amusement parks.

TIME REQUIRED

- 7 days (approximately 45 minutes each day; see Tables 3.7–3.8, pp. 37–38)

MATERIALS

- Classroom Map of the United States
- Safety glasses or goggles
- Computers with internet access
- Excel or other spreadsheet tool
- Rubber bands (latex-free)
- Plastic lids
- 1 cm × 1 cm square cross section of wood
- Wooden dowels
- Construction Paper
- Cardstock
- Kitchen towels
- Thumbtacks
- Various materials that can be used for Rube Goldberg machines (a variety of boards for ramps, marbles, old toys, blocks, string, and other items)
- Large gumdrops
- Smartphone or video camera
- Screen on which to play video, such as a computer or LCD projector

CONTENT STANDARDS AND KEY VOCABULARY

Table 4.5 lists the content standards from the *NGSS, CCSS,* and the Framework for 21st Century Learning that this lesson addresses, and Table 4.6 (p. 66) presents the key vocabulary. Vocabulary terms are provided for both teacher and student use. Teachers may choose to introduce some or all of the terms to students.

Table 4.5. Content Standards Addressed in STEM Road Map Module Lesson 2

NEXT GENERATION SCIENCE STANDARDS

PERFORMANCE EXPECTATIONS

- MS-PS3-1. Construct and interpret graphical displays of data to describe the relationships of kinetic energy to the mass of an object and to the speed of an object.

- MS-PS3-2. Develop a model to describe that when the arrangement of objects interacting at a distance changes, different amounts of potential energy are stored in the system.

- MS-PS3-4. Plan an investigation to determine the relationships among the energy transferred, the type of matter, the mass, and the change in the average kinetic energy of the particles as measured by the temperature of the sample.

- MS-PS3-5. Construct, use, and present arguments to support the claim that when the kinetic energy of an object changes, energy is transferred to or from the object.

SCIENCE AND ENGINEERING PRACTICES

Analyzing and Interpreting Data

Analyzing data in 6–8 builds on K–5 experiences and progresses to extending quantitative analysis to investigations, distinguishing between correlation and causation, and basic statistical techniques of data and error analysis.

- Construct and interpret graphical displays of data to identify linear and nonlinear relationships.

Developing and Using Models

Modeling in 6–8 builds on K–5 and progresses to developing, using and revising models to describe, test, and predict more abstract phenomena and design systems.

- Develop a model to describe unobservable mechanisms.

Planning and Carrying Out Investigations

Planning and carrying out investigations to answer questions or test solutions to problems in 6–8 builds on K–5 experiences and progresses to include investigations that use multiple variables and provide evidence to support explanations or design solutions.

- Plan an investigation individually and collaboratively, and in the design: identify independent and dependent variables and controls, what tools are needed to do the gathering, how measurements will be recorded, and how many data are needed to support a claim.

Engaging in Argument From Evidence

Engaging in argument from evidence in 6–8 builds on K–5 experiences and progresses to constructing a convincing argument that supports or refutes claims for either explanations or solutions about the natural and designed worlds.

- Construct, use, and present oral and written arguments supported by empirical evidence and scientific reasoning to support or refute an explanation or a model for a phenomenon.

Table 4.5. (*continued*)

DISCIPLINARY CORE IDEAS

PS3.A: Definitions of Energy

- Motion energy is properly called kinetic energy; it is proportional to the mass of the moving object and grows with the square of its speed.

- A system of objects may also contain stored (potential) energy, depending on their relative positions.

CROSSCUTTING CONCEPTS

Scale, Proportion, and Quantity

- Proportional relationships (e.g. speed as the ratio of distance traveled to time taken) among different types of quantities provide information about the magnitude of properties and processes.

Systems and System Models

- Models can be used to represent systems and their interactions—such as inputs, processes, and outputs—and energy and matter flows within systems.

Energy and Matter

- Energy may take different forms (e.g. energy in fields, thermal energy, energy of motion).

PS3.B: Conservation of Energy and Energy Transfer

- When the motion energy of an object changes, there is inevitably some other change in energy at the same time.

PS3.C: Relationship Between Energy and Forces

- When two objects interact, each one exerts a force on the other that can cause energy to be transferred to or from the object.

COMMON CORE STATE STANDARDS FOR MATHEMATICS

MATHEMATICAL PRACTICES

- MP1. Make sense of problems and persevere in solving them.
- MP2. Reason abstractly and quantitatively.
- MP3. Construct viable arguments and critique the reasoning of others.
- MP4. Model with mathematics.
- MP5. Use appropriate tools strategically.
- MP6. Attend to precision.

MATHEMATICAL CONTENT

- 6.SP.B.5b. Describing the nature of the attribute under investigation, including how it was measured and its units of measurement.

Table 4.5. (*continued*)

COMMON CORE STATE STANDARDS FOR ENGLISH LANGUAGE ARTS

READING STANDARDS

- RI.6.1. Cite textual evidence to support analysis of what the text says explicitly as well as inferences drawn from the text.

- RI.6.4. Determine the meaning of words and phrases as they are used in a text, including figurative, connotative, and technical meanings.

- RI.6.7. Integrate information presented in different media or formats (e.g., visually, quantitatively) as well as in words to develop a coherent understanding of a topic or issue.

WRITING STANDARDS

- W.6.1. Write arguments to support claims with clear reasons and relevant evidence.

- W.6.2. Write informative/explanatory texts to examine a topic and convey ideas, concepts, and information through the selection, organization, and analysis of relevant content.

SPEAKING AND LISTENING STANDARDS

- SL.6.1. Engage effectively in a range of collaborative discussions (one-on-one, in groups, and teacher-led) with diverse partners on grade 6 topics, texts, and issues, building on others' ideas and expressing their own clearly.

- SL.6.2. Interpret information presented in diverse media and formats (e.g., visually, quantitatively, orally) and explain how it contributes to a topic, text, or issue under study.

LITERACY STANDARDS

- L.6.1. Demonstrate command of the conventions of standard English grammar and usage when writing or speaking.

FRAMEWORK FOR 21ST CENTURY LEARNING

Creativity and Innovation, Critical Thinking and Problem Solving, Communication and Collaboration, Information Literacy, Media Literacy, ICT Literacy, Flexibility and Adaptability, Initiative and Self-Direction, Social and Cross Cultural Skills, Productivity and Accountability, Leadership and Responsibility, Economic, Business, and Entrepreneurial Literacy

---

Table 4.6. Key Vocabulary in Lesson 2

Key Vocabulary	Definition
cost-benefit analysis	the comparison of what an activity costs versus the outcomes or benefits of the activity
energy transfer	conversion of one type of energy to another, such as from mechanical energy to heat energy or from chemical energy to sound energy
friction	force that opposes motion and can cause a change in energy to heat or sound
heat energy	energy that changes the temperature of an object, typically caused by friction
kinetic energy	the energy of movement
mechanical energy	the energy an object has because of its position or motion
potential energy	the amount of stored energy in a physical situation (this could be due to the height of an object that has potential to be converted to kinetic energy because of gravity)
simple machines	the six basic mechanical devices that make up all other, complex machines: inclined plane, wedge, screw, lever, wheel and axle, and pulley
sound energy	energy that vibrates an object and causes it to make a sound, typically due to friction
sustainability	ability to maintain a rate of use of a resource that does not exceed the rate at which it is restored or created
total energy	the sum of kinetic and potential energy at any point in an object's trajectory; includes any thermal, electrical, sound, and light energy, among other types
trebuchet	a catapult device that uses a slingshot to launch an object a large distance; used throughout history as a weapon, in a variety of sizes and designs

TEACHER BACKGROUND INFORMATION

You should be familiar with the calculations for kinetic energy ($KE = \frac{1}{2} mv^2$) and potential energy ($PE = mgh$, where g is equal to 9.8 m/s^2 [32.2 ft/s^2]), know that total energy is equal to the sum of KE and PE (assuming that no energy is lost to friction), and understand that total energy remains constant for any closed system. You should also have a working knowledge of Excel or other spreadsheet tool. Before having your students work on their business plans, you may want to read this online article on how to motivate a group in writing a business proposal: *www.24hrco.com/images/articles/html/Pease_Mar13.html*.

COMMON MISCONCEPTIONS

Students will have various types of prior knowledge about the concepts introduced in this lesson. Table 4.7 outlines some common misconceptions students may have concerning these concepts. Because of the breadth of students' experiences, it is not possible to anticipate every misconception that students may bring as they approach this lesson. Incorrect or inaccurate prior understanding of concepts can influence student learning in the future, however, so it is important to be alert to misconceptions such as those presented in the table.

Table 4.7. Common Misconceptions About the Concepts in Lesson 2

Topic	Student Misconception	Explanation
Engineering design process (EDP)	Engineers use only the scientific method to solve problems in their work.	The scientific method is used to test predictions and explanations about the world. The EDP, on the other hand, is used to create a solution to a problem. In reality, engineers use both processes. (See pp. 9–11 for a discussion of this topic.)
Kinetic and potential energy (total energy)	Energy is lost in energy transformations.	The law of conservation of energy states that the amount of energy in an isolated system is conserved, or stays constant; however, friction may result in some small amounts of energy being transformed into sound or heat instead of into potential or kinetic energy.
	An object at rest has no energy.	An object at rest has potential energy if it is at a height above the ground.
	The terms *energy* and *force* are interchangeable.	Energy is related to mass and speed (in the case of kinetic energy) and position (in the case of potential energy) and is a scalar quantity. Force is a vector quantity and is an interaction between two objects.

PREPARATION FOR LESSON 2

Review the Teacher Background Information provided, and preview the websites suggested in the Learning Plan Components section below. Also preview all videos, checking for appropriate language and identifying points at which to stop the videos to explain concepts to students as needed. Gather the materials for the Rube Goldberg machines and for the trebuchet potential energy and kinetic energy investigation. If you do cannot supply enough materials for students to make Rube Goldberg machines, the activity can be adapted to be done with the videos provided within the lesson.

LEARNING PLAN COMPONENTS
Introductory Activity/Engagement

Connection to the Challenge: Begin each day of this lesson by directing students' attention to the driving question for the module and challenge: How can we use what we know about the development of amusements, the ways people experience thrills, and the laws of physics to propose new amusements that are both safe and extreme? Hold a brief student discussion of how their learning in the previous days' lessons has contributed to their ability to create their plan for their innovation in the final challenge. You may wish to create a class list of key ideas on chart paper or the board or have students create a STEM Research Notebook entry with this information.

To introduce this lesson, have students explore the superlatives of amusements by researching the fastest, highest, longest, and other extremes of amusements and games found at an amusement park. Then, have them research the energy sources for these rides and games. Ask them how they think energy might affect amusements. What might they consider in designing an amusement that is both thrilling and safe?

Science Class: Have each student choose one type of amusement ride or dart- or ball-throwing game, and then research the fastest, highest, longest drop, or other superlative associated with it. For example, students who choose roller coasters should look up the fastest, tallest, and steepest coasters (some suggested websites for this example are listed at the end of this section). Students who choose Ferris wheels may look up the highest, largest diameter, and most luxurious cabins. Students who choose games can examine the level of skill needed for games of different designs. Students should create data tables to record information about the rides or games, such as speed at the slowest point, speed at the fastest point, duration of ride, and height requirement.

Encourage students to make a wide variety of choices for their amusement rides or games, because in the challenge at the end of the module, groups will need experts on many different types of rides to work together to design a park incorporating a variety of elements. You can demonstrate to students how amusement parks have a variety of rides, such as Cedar Point (*www.cedarpoint.com*). (Note that under the Play tab on this website, you can find specific information on the rides, including speed, height, and duration. This information may be useful when students design their rides.)

World's fastest roller coasters

- *https://themysteriousworld.com/top-10-fastest-roller-coasters-in-the-world*

World's steepest roller coaster

- *http://themeparks.about.com/od/rollercoasterarticles/ss/steepest-roller-coasters.htm*

Tallest roller coaster

- *www.enki-village.com/top-10-tallest-roller-coasters.html*

Roller coaster superlatives

- *www.digitaltrends.com/cool-tech/biggest-rollercoasters-in-the-world*

Mathematics Connection: Begin by having students share the types of data they compiled in the science section of the lesson. Students who have similar data types should work together to decide the type of graph that are appropriate for each type of data, and then create the graph individually. For example, a bar graph would be best for a comparison of the top speeds achieved by different roller coasters, whereas a line graph would be more appropriate to plot the speeds over the duration of the ride. Students can do a gallery walk when graphs are complete to examine all the types of data and to ask questions. Then, have students use the Annenberg Learner Amusement Park Physics interactive website to design roller coaster features that create thrills but demonstrate safety issues: *www.learner.org/interactives/parkphysics/parkphysics.html*.

ELA Connection: Have students investigate different types of visual communication, such as diagrams, maps, graphs, and charts, and explain the pros and cons of each.

STEM Research Notebook Prompt

Ask students to interpret design features from National Geographic's JASON Digital Lab Coaster Creator website at *http://content3.jason.org/resource_content/content/digitallab/4859/misc_content/public/coaster.html*. Using this information, they should create a communication document for all design features of their ride or game in their STEM Research Notebooks. This document will be used later to support the development of the business plan.

Social Studies Connection: Have students map where amusement parks with the rides known for superlatives (from the prior Science Class) are located in the United States on a classroom map of the United States.

STEM Research Notebook Prompt

Ask students to answer the following questions in their Research Notebooks:

- Why do you think amusement parks are located in these areas?

- Do you see any patterns?

Activity/Exploration

In this portion of the lesson, students review simple machines, build Rube Goldberg machines, and analyze the energy transfers, noting extreme points of potential and kinetic energy. Students continue to build their knowledge for the challenge by examining costs, safety, and sustainability issues of amusement parks. Creating the physical Rube Goldberg machines and analyzing the energy transfer is intended to give students some ideas to make their rides creative and compelling.

Science Class: First, have students do a jigsaw activity to review the six types of simple machines and their uses. To perform this activity, divide students into six groups to become experts on the following topics (a website is suggested for each as a place to begin their research):

1. Levers (*www.sophia.org/tutorials/simple-machines-levers--3*)

2. Wedges (*www.sophia.org/tutorials/simple-machines-wedgeinclined-plane--2*)

3. Screws (*www.sophia.org/tutorials/simple-machines-screw*)

4. Inclined planes (*www.youtube.com/watch?v=NwDrp0UE3MM*)

5. Wheels and axles (*http://eschooltoday.com/science/simple-machines/what-is-a-wheel-and-axle.html*)

6. Pulleys(*www.khanacademy.org/science/physics/discoveries/simple-machines-explorations/a/pulleys*)

Give students 20 minutes to define their group's simple machine, find three examples of the simple machine in everyday use, show how the machine creates mechanical advantage when used, and create a short quiz about the main ideas from what they learned to give to their home group. Then, have students form home groups that each consist of one expert on inclined planes, one on levers, one on wedges, one on screws, one on wheels and axles, and one on pulleys. Students should take 10 minutes each to teach their group about their simple machine so that the whole group will come to understand all six simple machines. Students should then take each other's quizzes and go over them to check for understanding.

To review energy transfer (kinetic and potential energy), students can use the Energy Skate Park interactive simulation found at *https://phet.colorado.edu/en/simulation/legacy/energy-skate-park*. In this simulation, students can review conservation of energy by building tracks and ramps and examining the kinetic and potential energy as the skater moves. To check for understanding, have students present one of their track designs and describe the points of greatest kinetic and greatest potential energy and explain why. Also have them describe the points of lowest kinetic energy (should be the same as greatest potential energy) and lowest potential energy (should be the same as greatest kinetic energy).

Next, students will create Rube Goldberg machines, a contraption that performs a simple task in a very complicated way using a chain reaction of energy transfers (see *www.rubegoldberg.com/education*). First, have students watch the videos below for examples of Rube Goldberg machines. Because of the popularity of Rube Goldberg machine designs, a documentary has been made about them. It is called "Mousetrap to Mars," and the trailer can be viewed at *https://vimeo.com/3845644*. Both online and live Rube Goldberg machine competitions are held, and many universities have teams that compete at high levels. Purdue University has hosted one for over 20 years, and a video of Ferris State University's entry in this competition can be viewed at *https://vimeo.com/4687067*. Additional videos of Rube Goldberg machines can be found at *http://coolmaterial.com/roundup/rube-goldberg-machines*, and a Rube Goldberg Photobooth is featured at *https://vimeo.com/22111968*.

Then, have students form groups to build a Rube Goldberg machine that includes at least three different simple machines. Ask them to identify the following:

- Which simple machines are used
- Where mechanical energy converts to heat energy or sound energy
- The greatest and lowest points of potential energy and the greatest and lowest points of kinetic energy

If you cannot supply enough materials for this activity, you can have students identify the above while watching a video of a Rube Goldberg machine.

Mathematics Connection: Students can use the equations for potential energy (PE) and kinetic energy (KE) to calculate the PE, KE, and total energy for each element of the Rube Goldberg Machine (see Teacher Background Information on p. 66). Students should graph the quantities and be able to show that when KE rises, PE decreases, by graphing the horizontal position (starting point is $X = 0$) on the *x*-axis and by graphing KE in one color and PE in another color on a double-line graph. The sum of the two always equals the same amount of total energy, except when energy is lost to friction, heat, and sound. Students can follow up this activity by finding trends in the Energy Skate Park interactive simulations that show the changing PE and KE when going up and down a ramp:

https://phet.colorado.edu/en/simulation/energy-skate-park-basics and *https://phet.colorado.edu/en/simulation/legacy/energy-skate-park*.

ELA Connection: Ask students to consider what type of amusement they want to design for the park during the challenge. Then, tell them to write a narrative about what a person is feeling when he or she rides on the student-designed amusement. Students should not only describe the emotions of the rider but also give a sense of the physics of the ride. In this essay, students should explain the energy requirements of their ride: Does it require constant energy, such as a motor driving the ride, or is it similar to a roller coaster, which has a motor to take the cart to the highest point and then uses gravity for the downward portions of the ride?

Social Studies Connection: Tell students that Rube Goldberg is not the only person who was interested in the elaborate conversion of energy to do something simple. Many people around the world have done similar things in their cultures without knowing about the work of Rube Goldberg. Examine the work of people from different countries who have made similar creations.

- Australia—Bruce Petty (*http://trove.nla.gov.au/people/569273?c=people*)

- Austria—Franz Gsellmann (*www.weltmaschine.at*)

- United Kingdom—W. Heath Robinson: (*www.bbc.com/news/magazine-27813927*)

- India—Sukumar Ray (*http://thekahaniproject.org/khuror-kol-by-sukumar-ray-from-abol-tabol*)

STEM Research Notebook Prompt

Ask students the following questions about the fact that people are often interested in similar concepts, although they interpret them differently depending on the culture:

- Why do you think this type of work is done similarly around the world?

- Do you notice cultural differences even though it is similar work?

Explanation

In this section, students perform an investigation to calculate the kinetic and potential energy of a body flying a trajectory out of a trebuchet, a hurling device similar to a catapult or slingshot. Small groups of students make videos of the launch and calculate the kinetic and potential energy for five positions throughout the flight of an object. *Safety note:* Students should wear eye protection while using the trebuchet, as well as during activity setup and takedown. Remind students not to eat any food used in lab investigations, in this case the gumdrops.

Science Class: So that students get a better sense of how objects move in their designed amusements, have them study the motion of a trebuchet and the flung object. First, show them the following video of the Punkin Chunkin sport so that they will understand what a trebuchet does: *www.youtube.com/watch?v=qC6RJxFEMfY*. Then, conduct a whole-class discussion to identify the working parts of a trebuchet. Students should also watch the following video on how to build a model Roman catapult, or trebuchet: *www.youtube. com/watch?v=DwZA3WS2fB4*. If you are unable obtain the resources to make these model trebuchets, students can build one from paper, as demonstrated in this video: *www.youtube.com/watch?v=gjEME1HIsQ8*.

The goal of this class is to build trebuchets that launch a gumdrop in an elliptical trajectory that can be filmed and goes through a photogate to measure the final speed of the object at the end of the trajectory path. Students should work in small groups to make working trebuchets, using the cycles of an engineering design process listed below. This design procedure will not be assessed, but it is a good procedure to ensure that the designs are sound. Also post or hand out the engineering design process (EDP) diagram found at the end of the lesson.

1. Define the requirements and constraints

2. Learn about possible solutions

3. Plan for solutions

4. Try: design and build prototype

5. Evaluate and test prototype

6. Decide what worked and what did not

7. Communicate results

8. Redesign prototype and test

9. Evaluate redesign

Students can use the following materials to construct their trebuchet: rubber bands (latex-free), plastic lids, 1 cm × 1 cm square cross section of wood, wooden dowels, construction paper, cardstock, kitchen towels, and thumbtacks. *Safety note:* Tell students to use caution when handling the thumbtacks, as they can puncture the skin.

When making the video, students must include a clear picture of a 30 cm ruler in a vertical position next to landing site of the gumdrop so that they will be able to calculate scale in mathematics class. All data should be organized and recorded in the STEM Research Notebook. Facilitate a discussion to explain how the kinetic energy and

potential energy have changed, but the total energy stayed the same throughout the launch. During the discussion, check that students understand the concepts of scales, ratios, potential energy, and kinetic energy.

Mathematics Connection: Have students use the data from the video that they recorded in their STEM Research Notebooks to calculate the potential energy (PE = mgh) of the gumdrop at five points in the trajectory, as shown in the Figure 4.1. Students should measure the mass of the gumdrop, use 9.8 m/s² as g (the acceleration due to gravity), and measure the height from the starting point vertically (as seen at point 1) on their screens by pausing the video at each point and taking the measurements.

Figure 4.1. Motion of the Gum Drop Projected From the Trebuchet

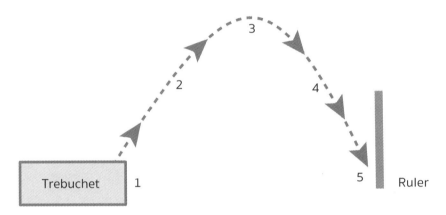

Once all measurements are taken from the video, students should use the ruler in the video to determine a ratio to convert their screen measurements to the actual measurements. The ratio is determined by dividing 30 cm by the length of the ruler as measured on the screen in centimeters. For example, if the rule on the screen measures 3 cm, the ratio used to convert the height measurements from the video screen to the actual measures is 10 (30 ÷ 3), so to obtain the actual measurements, students would then multiply the screen measurements by 10.

To find out if the scaled measures are accurate, students can compare their PE from point 3 with the KE measured with the photogate speed (final speed KE = ½ mv²). Ask students why the PE at point 3 should equal the KE at point 5 (refer to the Rube Goldberg calculations on p. 66 for a hint).

Elaboration/Application of Knowledge

Science Class: Have students research the safety concerns for amusement rides and for the park traffic flow in general. This knowledge will be the starting point for the development of the final report in the next lesson.

STEM Research Notebook Prompt

Students should create a matrix in their STEM Research Notebooks similar to the short example in Table 4.8 to record what they find in their research.

Table 4.8. Sample Matrix With a Few Entries

Safety Concern	Why It Is a Concern	What I Still Need to Know to Build My Park
Mechanical operations	Machines break down as a result of friction, so they must be inspected at regular intervals; ASTM International sets standards.	• How often do the inspections need to be done? • Whom do I contact to get inspectors to my park? • How much do the inspections cost?
Crowded conditions	More popular rides have longer lines and a higher density of crowds	• How do I account for the safety of people when I need to fit a large number of people in a small space?
Rider physical condition	To prevent legal issues, amusement parks need to post signs to warn people with medical conditions if they might not be safe on the ride.	• How do I personalize safety concerns without scaring away customers?

Mathematics Connection: Have students research how much it costs to build and run an amusement park and present their findings on an Excel or other spreadsheet. Begin with this video on how much it would cost to build a real-life Jurassic Park (stop the video at 3:00): *www.buzzfeed.com/alivelez/jurassic-park-would-cost-all-of-the-money#.ptAOwEpKz*. The following internet resources can be a starting point for student research:

- How to start an amusement park: *www.wikihow.com/Start-an-Amusement-Park*

- Academic paper with data on theme park development costs: *http://scholarworks. umass.edu/cgi/viewcontent.cgi?article=1072&context=gradconf_hospitality*

- Amusement park business plan: *www.bplans.com/amusement_park_business_plan/company_summary_fc.php*

- Q&A on a general budget for building an attraction: *www.themeparkinsider.com/news/response.cfm?id=2058*

STEM Research Notebook Prompt

Students should respond to the following questions in their Research Notebooks:

- What are the costs of operating an amusement park?

- How might you decide what to charge guests to be able to make money?

Students should conduct a peer review of their work to be sure that everyone has a common understanding of the costs involved in building and operating an amusement park.

ELA Connection: Have students explore ways to communicate the different aspects of design they must consider in the challenge—energy transfer, safety, sustainability, and cost-benefit analysis—in a business proposal that is two to three pages long. The following internet resources can help students organize the elements of their business plans:

- Sample amusement park business plan: *www.premieramusementdevelopers.com/sample-business-plan.html*

- Writing a business proposal: *http://smallbusiness.chron.com/steps-writing-business-proposal-70.html*

STEM Research Notebook Prompt

Ask students to respond to the following prompt in their Research Notebooks: If you were to start an amusement park business, what would you need to tell investors to convince them that your park would make money?

Social Studies Connection: Students now begin exploring "green" options in amusement park design. Have them work in small groups to develop executive summaries of green options that will make an amusement park more environmentally friendly. Explain to students that amusement parks use a great deal of energy, and recent improvements of amusement parks include attention to the carbon footprint of a park, or how much it is contributing to global climate change. Students can begin with the following resources and then expand their search:

- *www.mnn.com/family/family-activities/questions/do-green-amusement-or-theme-parks-exist*

- *www.entertainmentdesigner.com/news/theme-park-design-news/how-do-we-address-environmental-sustainability-in-entertainment-design*

- *www.greenisuniversal.com/learn/about-us/theme-parks*

- *www.mashable.com/2015/11/30/theme-park-solar-energy/#X5UVdT2t6Eq8*

STEM Research Notebook Prompt

Students should respond to the following prompt in their Research Notebooks: What can be done to make an amusement park more environmentally friendly, or green?

Evaluation/Assessment

Assessment of students occurs throughout the module in order to give students feedback on their work. The forms on pages 81–83 give ideas for ways you can assess student understanding in this module. Students may be assessed on the following measures:

- Rubric for Rube Goldberg Machine

- Scoring Guide for Engineering Design Process

- Scoring Guide for Kinetic and Potential Energy Calculations from Trebuchet

INTERNET RESOURCES

How to motivate a group in writing a business proposal
- *www.24hrco.com/images/articles/html/Pease_Mar13.html*

Cedar Point amusement park
- *www.cedarpoint.com*

World's fastest roller coasters
- *https://themysteriousworld.com/top-10-fastest-roller-coasters-in-the-world*

World's steepest roller coaster
- *http://themeparks.about.com/od/rollercoasterarticles/ss/steepest-roller-coasters.htm*

Tallest roller coaster
- *www.enki-village.com/top-10-tallest-roller-coasters.html*

Roller coaster superlatives
- *www.digitaltrends.com/cool-tech/biggest-rollercoasters-in-the-world*

Amusement Park Physics interactive website
- *www.learner.org/interactives/parkphysics/parkphysics.html*

JASON Digital Lab Coaster Creator
- *http://content3.jason.org/resource_content/content/digitallab/4859/misc_content/public/coaster.html*

Levers
- *www.sophia.org/tutorials/simple-machines-levers--3*

Wedges
- *www.sophia.org/tutorials/simple-machines-wedgeinclined-plane--2*

Screws
- *www.sophia.org/tutorials/simple-machines-screw*

Inclined planes
- *www.youtube.com/watch?v=NwDrp0UE3MM*

Wheel and axle
- *http://eschooltoday.com/science/simple-machines/what-is-a-wheel-and-axle.html*

Pulleys
- *www.khanacademy.org/science/physics/discoveries/simple-machines-explorations/a/pulleys*

"Mousetrap to Mars" trailer
- *https://vimeo.com/3845644*

Ferris State University's entry in a Rube Goldberg machine competition
- *https://vimeo.com/4687067*

11 Brilliant Rube Goldberg Machines
- *http://coolmaterial.com/roundup/rube-goldberg-machines*

Rube Goldberg Photobooth
- *https://vimeo.com/22111968*

Rube Goldberg Home Site
- *www.rubegoldberg.com*

Energy Skate Park interactive simulation
- *https://phet.colorado.edu/en/simulation/legacy/energy-skate-park*

- *https://phet.colorado.edu/en/simulation/energy-skate-park-basics*

Worldwide artists who have made creations similar to those of Rube Goldberg
- Australia—Bruce Petty: *http://trove.nla.gov.au/people/569273?c=people*

- Austria—Franz Gsellmann: *www.weltmaschine.at*

- United Kingdom —W. Heath Robinson: *www.bbc.com/news/magazine-27813927*

- India—Sukumar Ray: *http://thekahaniproject.org/khuror-kol-by-sukumar-ray-from-abol-tabol*

Punkin Chunkin trebuchet
- *www.youtube.com/watch?v=qC6RJxFEMfY*

How to build a model Roman catapult, or trebuchet
- *www.youtube.com/watch?v=DwZA3WS2fB4*

How to build a catapult, or trebuchet, out of paper
- *www.youtube.com/watch?v=gjEME1HIsQ8*

How much it would cost to build a real-life Jurassic Park
- *www.buzzfeed.com/alivelez/jurassic-park-would-cost-all-of-the-money#.ptAOwEpKz*

How to start an amusement park
- *www.wikihow.com/Start-an-Amusement-Park*

Academic paper with data on theme park development costs
- *http://scholarworks.umass.edu/cgi/viewcontent.cgi?article=1072&context=gradconf_hospitality*

Amusement park business plan
- *www.bplans.com/amusement_park_business_plan/company_summary_fc.php*

Q&A on a general budget for building an attraction
- *www.themeparkinsider.com/news/response.cfm?id=2058*

Sample amusement park business plan
- *www.premieramusementdevelopers.com/sample-business-plan.html*

Writing a business proposal
- *http://smallbusiness.chron.com/steps-writing-business-proposal-70.html*

Green amusement parks
- *www.mnn.com/family/family-activities/questions/do-green-amusement-or-theme-parks-exist*

- *http://entertainmentdesigner.com/news/theme-park-design-news/how-do-we-address-environmental-sustainability-in-entertainment-design*

- *www.greenisuniversal.com/learn/about-us/theme-parks*

RUBRIC FOR RUBE GOLDBERG MACHINE

Criteria	Emerging	Competent	Advanced
Machine must function to complete task safely.	• Does not accomplish a simple task	• Accomplishes a simple task with a simple set of steps	• Accomplishes a simple task with an extremely complex set of steps
Machine must include at least three different simple machines.	• Includes less than 3 different simple machines	• Includes 3 different simple machines	• Includes more than 3 different simple machines
Students must identify at least two positions when energy transfer happens from mechanical to heat or sound energy.	• Identifies less than 2 different positions where mechanical energy is transferred to sound or heat energy	• Identifies 2 different positions where mechanical energy is transferred to sound or heat energy	• Identifies more than 2 different positions where mechanical energy is transferred to sound or heat energy
Students must identify the highest and lowest points of both potential energy and kinetic energy.	• Do not identify highest and lowest points of both potential energy and kinetic energy	• Identifies highest and lowest points of both potential energy and kinetic energy	• Identifies the change in kinetic and potential energy throughout the duration of the machine operation and presents this information on a graph
Machine works as intended.	• Takes more than 3 tries to work as intended	• Takes 2 to 3 tries to work as intended	• Works as intended on the first try

COMMENTS:

SCORING GUIDE FOR ENGINEERING DESIGN PROCESS

Component	Emerging	Competent	Advanced
Design of trebuchet	• Student generates one design solution to the problem	• Student generates two or more alternate design solutions, listing the criteria for success	• Student generates a variety of design solutions and justifies solutions with comprehensive set of criteria for success
Methods of testing	• Some methods of testing are suggested	• Reasonable methods of testing are suggested for some of the listed criteria	• Reasonable methods for testing the trebuchet are described for each success criterion outlined
Planning	• Student produces a plan that contains some details of the required steps or resources	• Student produces a plan that contains a number of logical steps that include resources and time	• Student produces a plan that contains a number of detailed logical steps that describe the uses of resources and time

COMMENTS:

SCORING GUIDE FOR KINETIC AND POTENTIAL ENERGY CALCULATIONS FROM TREBUCHET

Activity	Below Expectations	Meets Expectations	Exceeds Expectations
Video has clear pictures of ruler and trajectory.			
Gumdrop final speed is measured by photogate.			
Scale measurements for height are accurate.			
Measurements from video are converted accurately into actual measurements.			
Calculation of PE from converted measurements is accurate.			
Calculation of KE from final velocity is accurate.			
Comparison of PE and KE is accurate.			

COMMENTS:

Lesson Plan 3: Amusement Park of the Future Design Challenge

This lesson is the culmination of the module and synthesizes all the knowledge and skills that students gained in the first two lessons. They have been conducting preliminary analyses and thinking about design features, and now they expand on them. Students work in small groups of four or five to create a final report for an innovative new park. Individual students are each responsible for an in-depth report on one amusement ride or game that will work harmoniously with those of other students in the group as part of a cohesive amusement park plan. Members of each group then consolidate their individual reports into a group report for the whole amusement park, making adaptations based on group suggestions.

ESSENTIAL QUESTION

- How do you collaboratively design an innovative amusement park while considering safety, impact on the community, costs, and marketing?

ESTABLISHED GOALS AND OBJECTIVES

At the conclusion of this lesson, students will be able to do the following:

- Write a report including blueprint, scale prototype drawing or mock-up, cost analysis, impact study, and marketing plan for their ride or game.

- Draw a scaled blueprint of their amusement park with their group members, taking into consideration foot traffic and refreshment issues.

- Conduct a cost-benefit analysis for their amusement park.

- Conduct an impact study for their proposed amusement park.

- Create a marketing plan and infomercial for their proposed amusement park.

TIME REQUIRED

- 13 days (approximately 45 minutes each day; see Tables 3.8–3.10, p. 38)

MATERIALS

- Computers with internet access

- Rulers for scale drawings

- Pencils for drawings

- Safety glasses or goggles

- Cardboard, paper, poster board, or balsa wood

- Glue and glue guns (be sure to caution students about the heat produced by glue guns or get the type that extrudes glue at a cool temperature)

- Large paper for drawing blueprints

- Excel or other spreadsheet tool

- Video camera or smartphone

- Editing software (optional)

CONTENT STANDARDS AND KEY VOCABULARY

Table 4.9 lists the content standards from the *NGSS, CCSS,* and the Framework for 21st Century Learning that this lesson addresses, and Table 4.10 (p. 88) presents the key vocabulary. Vocabulary terms are provided for both teacher and student use. Teachers may choose to introduce some or all of the terms to students.

Table 4.9. Content Standards Addressed in STEM Road Map Module Lesson 3

NEXT GENERATION SCIENCE STANDARDS

PERFORMANCE EXPECTATIONS

- MS-PS3-1. Construct and interpret graphical displays of data to describe the relationships of kinetic energy to the mass of an object and to the speed of an object.

- MS-PS3-2. Develop a model to describe that when the arrangement of objects interacting at a distance changes, different amounts of potential energy are stored in the system.

- MS-PS3-4. Plan an investigation to determine the relationships among the energy transferred, the type of matter, the mass, and the change in the average kinetic energy of the particles as measured by the temperature of the sample.

- MS-PS3-5. Construct, use, and present arguments to support the claim that when the kinetic energy of an object changes, energy is transferred to or from the object.

SCIENCE AND ENGINEERING PRACTICES

Analyzing and Interpreting Data

Analyzing data in 6–8 builds on K–5 experiences and progresses to extending quantitative analysis to investigations, distinguishing between correlation and causation, and basic statistical techniques of data and error analysis.

- Construct and interpret graphical displays of data to identify linear and nonlinear relationships

---

Table 4.9. (*continued*)

SCIENCE AND ENGINEERING PRACTICES (*continued*)

Developing and Using Models

Modeling in 6–8 builds on K–5 and progresses to developing, using, and revising models to describe, test, and predict more abstract phenomena and design systems.

- Develop a model to describe unobservable mechanisms.

Planning and Carrying Out Investigations

Planning and carrying out investigations to answer questions or test solutions to problems in 6–8 builds on K–5 experiences and progresses to include investigations that use multiple variables and provide evidence to support explanations or design solutions.

- Plan an investigation individually and collaboratively, and in the design: identify independent and dependent variables and controls, what tools are needed to do the gathering, how measurements will be recorded, and how many data are needed to support a claim.

Engaging in Argument From Evidence

Engaging in argument from evidence in 6–8 builds on K–5 experiences and progresses to constructing a convincing argument that supports or refutes claims for either explanations or solutions about the natural and designed worlds.

- Construct, use, and present oral and written arguments supported by empirical evidence and scientific reasoning to support or refute an explanation or a model for a phenomenon.

DISCIPLINARY CORE IDEAS

PS3.A: Definitions of Energy

- Motion energy is properly called kinetic energy; it is proportional to the mass of the moving object and grows with the square of its speed.

- A system of objects may also contain stored (potential) energy, depending on the relative positions of the objects.

PS3.B: Conservation of Energy and Energy Transfer

- When the motion energy of an object changes, there is inevitably some other change in energy at the same time.

PS3.C: Relationship Between Energy and Forces

- When two objects interact, each one exerts a force on the other that can cause energy to be transferred to or from the object.

CROSSCUTTING CONCEPTS

Scale, Proportion, and Quantity

- Proportional relationships (e.g. speed as the ratio of distance traveled to time taken) among different types of quantities provide information about the magnitude of properties and processes.

Table 4.9. (*continued*)

Systems and System Models
- Models can be used to represent systems and their interactions—such as inputs, processes, and outputs—and energy and matter flows within systems.

Energy and Matter
- Energy may take different forms (e.g. energy in fields, thermal energy, energy of motion).

COMMON CORE STATE STANDARDS FOR MATHEMATICS

MATHEMATICAL PRACTICES
- MP1. Make sense of problems and persevere in solving them.
- MP2. Reason abstractly and quantitatively.
- MP3. Construct viable arguments and critique the reasoning of others.
- MP4. Model with mathematics.
- MP5. Use appropriate tools strategically.
- MP6. Attend to precision.

MATHEMATICAL CONTENT
- 6.SP.B.5b. Describing the nature of the attribute under investigation, including how it was measured and its units of measurement.

COMMON CORE STATE STANDARDS FOR ENGLISH LANGUAGE ARTS

READING STANDARDS
- RI.6.1. Cite textual evidence to support analysis of what the text says explicitly as well as inferences drawn from the text.
- RI.6.4. Determine the meaning of words and phrases as they are used in a text, including figurative, connotative, and technical meanings.
- RI.6.7. Integrate information presented in different media or formats (e.g., visually, quantitatively) as well as in words to develop a coherent understanding of a topic or issue.

WRITING STANDARDS
- W.6.1. Write arguments to support claims with clear reasons and relevant evidence.
- W.6.2. Write informative/explanatory texts to examine a topic and convey ideas, concepts, and information through the selection, organization, and analysis of relevant content.

SPEAKING AND LISTENING STANDARDS
- SL.6.1. Engage effectively in a range of collaborative discussions (one-on-one, in groups, and teacher-led) with diverse partners on grade 6 topics, texts, and issues, building on others' ideas and expressing their own clearly.

Table 4.9. (*continued*)

> **SPEAKING AND LISTENING STANDARDS** (*continued*)
> * SL.6.2. Interpret information presented in diverse media and formats (e.g., visually, quantitatively, orally) and explain how it contributes to a topic, text, or issue under study.
> * SL.6.5. Include multimedia components (e.g., graphics, images, music, sound) and visual displays in presentations to clarify information.
>
> **LITERACY STANDARDS**
> * L.6.1. Demonstrate command of the conventions of standard English grammar and usage when writing or speaking.
>
> **FRAMEWORK FOR 21ST CENTURY LEARNING**
> Creativity and Innovation, Critical Thinking and Problem Solving, Communication and Collaboration, Information Literacy, Media Literacy, ICT Literacy, Flexibility and Adaptability, Initiative and Self-Direction, Social and Cross Cultural Skills, Productivity and Accountability, Leadership and Responsibility, Economic, Business, and Entrepreneurial Literacy

Table 4.10. Key Vocabulary in Lesson 3

Key Vocabulary	Definition
blueprint	a technical drawing that is done to describe the actual object and location using a scale
impact study	research done on a topic to determine how a certain action (such as building an amusement park) will affect phenomena such as traffic, wildlife, and tree growth
prototype	a model mock-up of an object

TEACHER BACKGROUND INFORMATION

Although blueprints have been replaced by whiteprints and, more recently, computer-aided design and other technologies, the term *blueprint* is still used informally for all such images. In this lesson, students create traditional blueprint drawings for their amusement parks. Typical scales for blueprints are 1:20, 1:50, and 1:100, and they are usually drawn in metric units. You should be familiar with some basic amusement industry ideas, such as placing the most popular ride toward the back of the park so that patrons must go past other rides that might catch their attention and trying to reduce the appearance of long lines by having them zigzag back and forth.

COMMON MISCONCEPTIONS

Students will have various types of prior knowledge about the concepts in this lesson. Table 4.11 outlines some common misconceptions students may have concerning these concepts. Because of the breadth of students' experiences, it is not possible to anticipate every misconception that students may bring as they approach this lesson. Incorrect or inaccurate prior understanding of concepts can influence student learning in the future, however, so it is important to be alert to misconceptions such as those presented in the table.

Table 4.11. Common Misconceptions About the Concepts in Lesson 3

Topic	Student Misconception	Explanation
Marketing	More marketing is always better.	The most effective marketing strategies tend to be very focused and high quality.
	Any content counts as marketing.	Marketing needs to be a unified campaign with consistent messaging and branding.

PREPARATION FOR LESSON 3

Review the Teacher Background Information provided, and preview the websites suggested in the Learning Plan Components section below. In this lesson, it is critical to form groups of four or five students in which each member contributes an original, distinctly different amusement, but with the entire group having a common theme for the whole park. In this way, groups will be well balanced and able to create a comprehensive park when members consolidate their amusements. For instance, if a group wants to create a water park, each of the members should have a different idea for one amusement ride or game; these components should collectively create an entire water park with four or five rides or games. Alternatively, a group may choose to put together a roller coaster park, with each student in the group having a roller coaster as his or her amusement, but each roller coaster should have a different emphasis. Another important task is to monitor the way groups report their business plans. Individual students' business plans will be different, but they should keep in mind that these plans will need to be synthesized into a single plan for the group. Your role will be primarily to monitor individual and group progress and answer questions.

LEARNING PLAN COMPONENTS
Introductory Activity/Engagement

Connection to the Challenge: Begin each day of this lesson by directing students' attention to the driving question for the module and challenge: How can we use what we know about the development of amusements, the ways people experience thrills, and the laws of physics to propose new amusements that are both safe and extreme? Ask students what they would include if they were to design an amusement park, and why. Hold a brief student discussion of how their learning in the previous days' lessons contributed to their ability to create their plan for their innovation in the final challenge. Students should review all their notes in the STEM Research Notebook for this task. You may wish to create a class list of key ideas on chart paper or the board or have students create a STEM Research Notebook entry with this information.

Science Class: Tell students that they will choose an amusement park that they most want to emulate for their projects. Based on their decisions, they will then form groups. The following website offers maps of various amusement parks for inspiration: *http://www.themeparkbrochures.net/maps*.

STEM Research Notebook Prompt

Students should respond to the following questions in their Research Notebooks:

- Examine the layout of your chosen park on a map. How are the rides placed in the overall park? Where are the refreshments placed?

- Examine where the park is located in relation to nearby cities and towns. Why do you think the park is located there? Who might be going to the park?

- Find the most extreme ride. Where is it located in the park? Why do you think it was placed there? What is the ride focused on in terms of physical thrills? Why it is designed that way based on KE and PE changes?

- Note that there are maps of the same park for different years. Explain how your chosen park has changed over time. What might you change for your park in the future?

Mathematics Connection: Have students consider some mathematical questions regarding the amusement park that they chose in science class as the one they most want to emulate for their projects.

STEM Research Notebook Prompt

Have students answer the following questions in their Research Notebooks:

- Examine the map of your chosen park. What is the scale of the map compared with the actual size of the park and rides?

- How much space is in between each ride? Why do you think this is the case?

- How are the parks arranged in terms of people forming lines? Why do they have this arrangement?

- Do you notice any other things about the park arrangement that are important? What are they?

ELA Connection: Read about environmental impact studies for the following projects from the Executive Summaries:

- Keystone XL Project: *https://2012-keystonepipeline-xl.state.gov/documents/ organization/181185.pdf*

- Construction of a new middle school in New York State: *www.ccsd.ws/district. cfm?subpage=963265*

STEM Research Notebook Prompt

Students should choose one impact study from the examples and answer the following questions in their STEM Research Notebooks, then discuss as a whole class:

- What do you think is the purpose of and Environmental Impact Assessment?

- What environmental impacts are relevant for your amusement park? What questions should be asked about the impact of building an amusement park on the environment?

- Are there other questions you need to answer to plan about the land and environment for your amusement park?

Social Studies Connection: Have students read the following article with predictions for the future of amusement parks: *http://mentalfloss.com/article/64377/5-educated-predictions-future-amusement-parks.*

STEM Research Notebook Prompt

Ask students to answer the following questions in their Research Notebooks:

- Which of the five predictions do you think are most important for your planned park? Least important? Why?

Then, have students consider the theme they want for their parks and ask them to respond to the following questions in their notebooks:

- Why do you want your park to have that theme?

- Who will be the audience for your park? Is it going to attract people who like many different kinds of amusements or focus on one type?

Activity/Exploration

Science Class: Students work individually to build scale models, or prototypes, of their amusement rides or games. They can practice their skills at creating a scale drawing by first drawing the dimensions of the classroom at a 1:20 scale. Once students have mastered their ability to create scale drawings, they should proceed to convert their ride or game design to a 1:20 scale. The purpose of this prototype is to illustrate the physical features of the ride or game. This prototype will not be an operating version, but rather will be similar to the scale models that architects use to describe the features of their designs. For their prototypes, students can use simple materials such as cardboard, paper, poster board, or balsa wood, along with glue and glue guns. *Safety notes*: Be sure to have students wear eye protection while building their models. Tell them to use caution when working with the glue guns, which get hot and can burn skin and potentially cause fire. If you are unable to provide adequate supplies for building models, students can create drawings that show the rides or games from multiple perspectives.

Mathematics Connection: Have students write and present a report to accompany the prototype. The organization of the individual reports should mirror the organization of the group report. Each report should have the following sections:

- Blueprint of ride or game

- Scale prototype drawing or mock-up

- Cost analysis

- Impact study

- Marketing plan

Give students the Rubric for Presentation and Report (pp. 96–98) so that they will understand what they need to consider. Then, have students complete the following mathematics-related sections of the report for their prototype:

- Scale of prototype

- Source of energy

- Types of energy transfer

- Kinetic and potential energy graphs for the duration of the ride or game

- Cost to build this type of ride

- Cost to maintain this type of ride

Social Studies Connection: Students continue writing their reports to accompany the prototype. In social studies class, have students complete the following sections of the report and presentation on their prototype:

- History of this type of ride or game

- Audience for ride or game

- Features providing thrills to the audience

- Psychological reasons people experience thrills

ELA Connection: Now, have students present their prototypes to the class along with their reports. Ask the entire class to conduct a peer review of the presentations and reports. The following protocol should be used for the peer review:

1. A student gives his or her presentation.

2. Peer reviewers ask clarifying questions, and the presenting student responds.

3. Peer reviewers explain all positive responses to the prototype by starting each response with the phrase "I like …," while the presenting student listens and takes notes.

4. Peer reviewers explain all responses regarding potential improvements to the prototype by starting each response with the phrase "I wish …," while the presenting student listens and takes notes.

5. The presenting student responds to the peer review by summarizing his or her next steps in redesigning the prototype.

Explanation

Science Class: Based on students' answers to the questions earlier about their interests, break the students into groups of four or five who have similar ideas for themes but different amusement rides and games (see preparation for Lesson 3 starting on p. 84 for guidelines on forming these groups). Ask them to discuss with their groups until the group members come to a consensus about the theme and audience for their park. Then, have each group tell you its final decision.

Students should add to their reports based on the feedback received during the peer review, including the positive aspects of the amusement, the improvements suggested, and the improvements actually incorporated into the prototype. Students also should revise or rebuild their prototypes based on the peer review. Once their reports and prototypes have been updated, students combine their individual reports into an integrated group report. Students also work on a group presentation for the overall project, referring to the rubric for the necessary parts of the presentation.

Mathematics Connection: Have group members collaboratively draw a blueprint of their park and add refreshments, parking, areas for lines, and restrooms. Students should include the scale of the drawing and appropriate labels on their blueprints.

ELA Connection: Now, students will work in their amusement park groups to create a marketing plan for their proposed park. Students will conduct research to be able to construct an additional section of the report and should include the following:

- Location of the park and why this location was selected

- Information on the customers, including demographics, needs, and buying decisions

- Description of how the proposed park is innovative compared with existing parks

- List of advertising outlets: TV, radio, internet, print (students should look at advertisements for amusement parks for inspiration)

- Mission statement, including whom students are selling to, what they are selling, and their unique selling propositions

- Price for admission

- Explanation of how they will monitor their results (e.g., following up with customer satisfaction surveys, monitoring costs as a percentage of sales)

Also have each group create a video commercial providing key information about its park and using group members' prototypes and blueprints in the video.

Social Studies Connection: Have students conduct research for an environmental impact assessment for the location of their park. The purpose of this assessment is to be sure that students minimize environmental damage when building their parks. This should be another new section of the report and should include the following:

- Area needed for park, including adequate parking for projected number of people entering park (should correspond to projected customers in the business plan)

- Number of trees to be cut down (if applicable)

- Types of wildlife that live in the area and how they might be affected

- Types of plants that grow in the area and how they might be affected

Elaboration/Application of Knowledge

Science Class, Mathematics and ELA Connections, and Social Studies Connection: This module ends with groups giving multimedia presentations to an audience of teachers, other students, and invited business people from the community. The presentation should include the group's report, blueprint with explanation, prototype, cost analysis, impact study, marketing plan, and video commercial.

Ask audience members to assess the presentations using the Rubric for Presentation and Report (pp. 96–98) and provide verbal comments to the presenting group. The presenting group should reflect the feedback to the audience to ensure a common understanding of the feedback.

Evaluation/Assessment

Students will need feedback on the quality of the group blueprints and business plans. Students may be assessed on the Rubric for Presentation and Report.

INTERNET RESOURCES

Maps of amusement parks around the world
- *www.themeparkbrochures.net/maps*

Summary of environmental impact studies
- *https://en.wikipedia.org/wiki/Environmental_impact_assessment*

Predictions for amusement parks of the future
- *http://mentalfloss.com/article/64377/5-educated-predictions-future-amusement-parks*

RUBRIC FOR PRESENTATION AND REPORT

Component	Emerging	Competent	Advanced
Blueprint	Blueprint is not to scale or is lacking any of the following elements: • Appropriate labels • Refreshments • Parking • Areas for lines • Restrooms	Blueprint is to scale and has appropriate labels. Incorporates refreshments, parking, areas for lines, and restrooms.	Blueprint is to scale and has appropriate labels. Incorporates refreshments, parking, areas for lines, and restrooms, as well as other theme-related rides or extra features.
Scale prototype	Prototype is not to scale or is missing or inaccurately represents any of the following: • Source of energy • Types of energy transfer • Kinetic and potential energy graphs for duration of the ride or game	Prototype is to scale and accurately represents all of the following: • Source of energy • Types of energy transfer • Kinetic and potential energy graphs for duration of the ride or game	Prototype is to scale, has all required elements needed for competent performance, and includes additional details.
Cost analysis	One or both of the following items are missing or inaccurate in the report: • Cost to build this type of ride • Cost to maintain this type of ride	Both of these items are included in the report accurately: • Cost to build this type of ride • Cost to maintain this type of ride	All the following items are included in the report accurately: • Cost to build this type of ride • Cost to maintain this type of ride • Additional details related to the cost of building and maintaining the ride

Rubric for Presentation and Report (*continued*)

Component	Emerging	Competent	Advanced
Impact study	One or more of the following items are missing or inaccurate in the report: • Area needed for park, including adequate parking for projected number of people entering park (should correspond to projected customers in business plan) • Number of trees to be cut down (if applicable) • Types of wildlife that live in the area and how they might be affected • Types of plants that grow in the area and how they might be affected	All the following items are included in the report accurately: • Area needed for park, including adequate parking for projected number of people entering park (should correspond to projected customers in business plan) • Number of trees to be cut down (if applicable) • Types of wildlife that live in the area and how they might be affected • Types of plants that grow in the area and how they might be affected	All the following items are included in the report accurately: • Area needed for park, including adequate parking for projected number of people entering park (should correspond to projected customers in business plan) • Number of trees to be cut down (if applicable) • Types of wildlife that live in the area and how they might be affected • Types of plants that grow in the area and how they might be affected • Additional reasonable predictions about environmental impact of the park

Rubric for Presentation and Report (*continued*)

Component	Emerging	Competent	Advanced
Marketing plan	One or more of the following items are missing or inaccurate in the report: • Location of the park and why selected • Customers • How the proposed park is innovative compared with other parks • Advertising outlets • Mission statement • Cost of admission • How results will be monitored	All the following items are included in the report accurately: • Location of the park and why selected • Customers • How the proposed park is innovative compared with other parks • Advertising outlets • Mission statement • Cost of admission • How results will be monitored	All the following items are included in the report accurately: • Location of the park and why selected • Customers • How the proposed park is innovative compared with other parks • Advertising outlets • Mission statement • Cost of admission • How results will be monitored • Additional details about marketing in both the report and presentation
Video commercial	Video commercial is difficult to understand or does not present the amusement park favorably.	Video commercial is clear and presents the amusement park favorably.	Video commercial is clear and presents the amusement park in an extremely compelling way.
Feedback	Student interpretation of feedback given at presentation is not accurate.	Student interpretation of feedback given at presentation is accurate.	Student interpretation of feedback given at presentation is accurate and a plan of action for another cycle of design is offered

COMMENTS:

5

TRANSFORMING LEARNING WITH AMUSEMENT PARK OF THE FUTURE AND THE *STEM ROAD MAP CURRICULUM SERIES*

Carla C. Johnson

This chapter serves as a conclusion to the Amusement Park of the Future integrated STEM curriculum module, but it is just the beginning of the transformation of your classroom that is possible through use of the *STEM Road Map Curriculum Series*. In this book, many key resources have been provided to make learning meaningful for your students through integration of science, technology, engineering, and mathematics, as well as social studies and English language arts, into powerful problem- and project-based instruction. First, the Amusement Park of the Future curriculum is grounded in the latest theory of learning for children in elementary school specifically. Second, as your students work through this module, they engage in using the engineering design process (EDP) and build prototypes like engineers and STEM professionals in the real world. Third, students acquire important knowledge and skills grounded in national academic standards in mathematics, English language arts, science, and 21st century skills that will enable their learning to be deeper, retained longer, and applied throughout, illustrating the critical connections within and across disciplines. Finally, authentic formative assessments, including strategies for differentiation and addressing misconceptions, are embedded within the curriculum activities.

The Amusement Park of the Future curriculum in the Innovation and Progress STEM Road Map theme can be used in single-content middle school classrooms (e.g., science) where there is only one teacher or expanded to include multiple teachers and content areas across classrooms. Through the exploration of the Amusement Park Challenge,

students engage in a real-world STEM problem on the first day of instruction and gather necessary knowledge and skills along the way in the context of solving the problem.

The other topics in the *STEM Road Map Curriculum Series* are designed in a similar manner, and NSTA Press plans to publish additional volumes in this series for this and other grade levels. The tentative list of other books includes the following themes and subjects:

- Innovation and Progress
  - Construction materials
  - Harnessing solar energy
  - Transportation in the future
  - Wind energy
- The Represented World
  - Rainwater analysis
  - Recreational STEM: Swing set makeover
- Sustainable Systems
  - Hydropower efficiency
  - Composting: reduce, reuse, recycle
- Optimizing the Human Condition
  - Water conservation: Think global, act local

If you are interested in professional development opportunities focused on the STEM Road Map specifically or integrated STEM or STEM programs and schools overall, contact the lead editor of this project, Dr. Carla C. Johnson (*carlacjohnson@purdue.edu*), associate dean and professor of science education at Purdue University. Someone from the team will be in touch to design a program that will meet your individual, school, or district needs.

APPENDIX

CONTENT STANDARDS ADDRESSED IN THIS MODULE

NEXT GENERATION SCIENCE STANDARDS

Table A1 (p. 102) lists the science and engineering practices, disciplinary core ideas, and crosscutting concepts this module adresses. The supported performance expectations are as follows:

- MS-PS3-1. Construct and interpret graphical displays of data to describe the relationships of kinetic energy to the mass of an object and to the speed of an object.

- MS-PS3-2. Develop a model to describe that when the arrangement of objects interacting at a distance changes, different amounts of potential energy are stored in the system.

- MS-PS3-4. Plan an investigation to determine the relationships among the energy transferred, the type of matter, the mass, and the change in the average kinetic energy of the particles as measured by the temperature of the sample.

- MS-PS3-5. Construct, use, and present arguments to support the claim that when the kinetic energy of an object changes, energy is transferred to or from the object.

Table A1. *Next Generation Science Standards (NGSS)*

Science and Engineering Practices

ANALYZING AND INTERPRETING DATA

Analyzing data in 6–8 builds on K–5 experiences and progresses to extending quantitative analysis to investigations, distinguishing between correlation and causation, and basic statistical techniques of data and error analysis.

- Construct and interpret graphical displays of data to identify linear and nonlinear relationships.

DEVELOPING AND USING MODELS

Modeling in 6–8 builds on K–5 and progresses to developing, using and revising models to describe, test, and predict more abstract phenomena and design systems.

- Develop a model to describe unobservable mechanisms.

PLANNING AND CARRYING OUT INVESTIGATIONS

Planning and carrying out investigations to answer questions or test solutions to problems in 6–8 builds on K–5 experiences and progresses to include investigations that use multiple variables and provide evidence to support explanations or design solutions.

- Plan an investigation individually and collaboratively, and in the design: identify independent and dependent variables and controls, what tools are needed to do the gathering, how measurements will be recorded, and how many data are needed to support a claim.

ENGAGING IN ARGUMENT FROM EVIDENCE

Engaging in argument from evidence in 6–8 builds on K–5 experiences and progresses to constructing a convincing argument that supports or refutes claims for either explanations or solutions about the natural and designed worlds.

- Construct, use, and present oral and written arguments supported by empirical evidence and scientific reasoning to support or refute an explanation or a model for a phenomenon.

Disciplinary Core Ideas

PS3.A: DEFINITIONS OF ENERGY

- Motion energy is properly called kinetic energy; it is proportional to the mass of the moving object and grows with the square of its speed.

- A system of objects may also contain stored (potential) energy, depending on their relative positions.

PS3.B: CONSERVATION OF ENERGY AND ENERGY TRANSFER

- When the motion energy of an object changes, there is inevitably some other change in energy at the same time.

PS3.C: RELATIONSHIP BETWEEN ENERGY AND FORCES

- When two objects interact, each one exerts a force on the other that can cause energy to be transferred to or from the object.

Table A1. (*continued*)

Crosscutting Concepts

SCALE, PROPORTION, AND QUANTITY

- Proportional relationships (e.g. speed as the ratio of distance traveled to time taken) among different types of quantities provide information about the magnitude of properties and processes.

SYSTEMS AND SYSTEM MODELS

- Models can be used to represent systems and their interactions—such as inputs, processes, and outputs—and energy and matter flows within systems.

ENERGY AND MATTER

- Energy may take different forms (e.g. energy in fields, thermal energy, energy of motion).

Table A2. Common Core Mathematics and English Language Arts (ELA) Standards

MATHEMATICAL PRACTICES	READING STANDARDS
- MP1. Make sense of problems and persevere in solving them. - MP2. Reason abstractly and quantitatively. - MP3. Construct viable arguments and critique the reasoning of others. - MP4. Model with mathematics. - MP5. Use appropriate tools strategically. - MP6. Attend to precision. - MP8. Look for and express regularity in repeated reasoning. **MATHEMATICAL CONTENT** - 6.SP.B.5b. Describing the nature of the attribute under investigation, including how it was measured and its units of measurement.	- RI.6.1. Cite textual evidence to support analysis of what the text says explicitly as well as inferences drawn from the text. - RI.6.7. Integrate information presented in different media or formats (e.g., visually, quantitatively) as well as in words to develop a coherent understanding of a topic or issue. **WRITING STANDARDS** - W.6.1. Write arguments to support claims with clear reasons and relevant evidence. - W.6.2. Write informative/explanatory texts to examine a topic and convey ideas, concepts, and information through the selection, organization, and analysis of relevant content. **SPEAKING AND LISTENING STANDARDS** - SL.6.1. Engage effectively in a range of collaborative discussions (one-on-one, in groups, and teacher-led) with diverse partners on grade 6 topics, texts, and issues, building on others' ideas and expressing their own clearly. - SL.6.5. Include multimedia components (e.g., graphics, images, music, sound) and visual displays in presentations to clarify information. **LITERACY STANDARDS** - L.6.1. Demonstrate command of the conventions of standard English grammar and usage when writing or speaking.

Table A3. 21st Century Skills From the Framework for 21st Century Learning

21st Century Skills	Learning Skills and Technology Tools	Teaching Strategies	Evidence of Success
INTERDISCIPLINARY THEMES	• Economic, Business, and Entrepreneurial Literacy	• Business plan basics are introduced to support students in knowing how to make appropriate economic choices. • Students learn entrepreneurial skills to enhance workplace productivity and career options.	• Students present their blueprints, scale prototypes, cost-benefit analyses, impact studies, marketing plans, and infomercials to the class, teacher, and community members in the challenge.
LEARNING AND INNOVATION SKILLS	• Creativity and Innovation • Critical Thinking and Problem Solving • Communication and Collaboration	• Teachers have students use a wide range of idea creation techniques (such as brainstorming) and elaborate, refine, analyze and evaluate their own ideas to improve and maximize creative efforts. • Teachers create learning environments that help students demonstrate originality and inventiveness in work and understand the real-world limits to adopting new ideas. • Teachers help students accept failure as an opportunity to learn and understand that creativity and innovation are a long-term, cyclical process of small successes and frequent mistakes.	• Students create their business plans with a variety of modalities of learning and perspectives. • Students receive teacher and peer feedback in a formative way to inform the products that they develop throughout the problem- and project-based learning (PBL) module.

Table A3. (*continued*)

21st Century Skills	Learning Skills and Technology Tools	Teaching Strategies	Evidence of Success
INFORMATION, MEDIA AND TECHNOLOGY SKILLS	• Information Literacy • Media Literacy • ICT Literacy	• Teachers give students strategies to access information efficiently (time) and effectively (sources). • Students have opportunities to practice using information accurately and creatively for the issue or problem at hand.	• Students find reliable and relevant resources (both print and electronic) to build and integrate their knowledge about the design of amusement parks, physics, history, psychology, and engineering.
LIFE AND CAREER SKILLS	• Flexibility and Adaptability • Initiative and Self-Direction • Social and Cross Cultural Skills • Productivity and Accountability • Leadership and Responsibility	• Teachers give students strategies that help them work effectively in a climate of ambiguity and changing priorities and incorporate feedback effectively. • Teachers ask students to set goals with tangible and intangible success criteria, which leads to using time and managing workload efficiently.	• Students have to work collaboratively in groups as well as be able to integrate their independent work into a group project. • Students have to set benchmarks for accomplishing deliverables to reach the goals set by the PBL module.

Table A4. English Language Development Standards, Grades 6–8

ELD STANDARD 1: SOCIAL AND INSTRUCTIONAL LANGUAGE English language learners communicate for Social and Instructional purposes within the school setting.
ELD STANDARD 2: THE LANGUAGE OF LANGUAGE ARTS English language learners communicate information, ideas, and concepts necessary for academic success in the content area of Language Arts.
ELD STANDARD 3: THE LANGUAGE OF MATHEMATICS English language learners communicate information, ideas, and concepts necessary for academic success in the content area of Mathematics.
ELD STANDARD 4: THE LANGUAGE OF SCIENCE English language learners communicate information, ideas, and concepts necessary for academic success in the content area of Science.
ELD STANDARD 5: THE LANGUAGE OF SOCIAL STUDIES English language learners communicate information, ideas, and concepts necessary for academic success in the content area of Social Studies.

Source: World-Class Instructional Design and Assessment Consortium (WIDA), 2012, 2012 Amplification of the English language development standards: Kindergarten–grade 12, *www.wida. us/standards/eld.aspx.*

INDEX

Page numbers printed in **boldface type** indicate tables and figures.

L

learning cycle, 11–12

lesson plans. *See* Amusement Park of the Future Design Challenge lesson plan; Faster, Higher, and Safer lesson plan; History and Psychology of Amusement Parks lesson plan

M

marketing plan, Amusement Park of the Future Design Challenge lesson plan, **98**

marketing, STEM misconceptions, **89**

mass, 44

mathematics connections

    Amusement Park of the Future Design Challenge lesson plan, 90, 92–93, 94, 95

    Faster, Higher, and Safer lesson plan, 69, 71–72, 74, **74**, 75–76

    History and Psychology of Amusement Parks lesson plan, 47, 50, 52, 53

mechanical energy, 66

module launch, 26

module timeline, 36, **37–38**

N

Newtonian mechanics, 51

Next Generation Science Standards (*NGSS*)

    Amusement Park of the Future Design Challenge lesson plan, **85–87**

    described, 2, 100–101, **102–103**

    Faster, Higher, and Safer lesson plan, **63–64**

    and formative assessments, 15

    History and Psychology of Amusement Parks lesson plan, 42, **43**

O

optimizing the human experience theme, 5

P

potential energy, 66, **67**, 71, 72–74, **81**, **83**, 90, 93

problem scoping, engineering design process (EDP), 9–10, **10**

project- and problem-based learning, 9

    STEM Road Map Curriculum Series, 9

prototypes

    defined, **88**

    engineering design process (EDP), 11

    scale prototype, 84, 92, **96**

psychology, 44

R

the represented world theme, 4